PERCH

to Lorna Elizabeth

PERCH

Contemporary Days and Ways

John Bailey & Roger Miller

The Crowood Press

First published 1989 by
The Crowood Press
Ramsbury, Marlborough,
Wiltshire SN8 2HR

www.crowood.com

This impression 2007

British Library Cataloguing in Publication Data

Bailey, John
Perch: contemporary days & ways.
1. Perch. Angling – Manuals
I. Title   II. Miller, Roger
799.1'758
ISBN 978-1-85223-121-7

Acknowledgements

With sincere thanks to Vic Bellars, Steve Burke, John Nunn, Peter
Rodgers and Nigel Witham for their invaluable help.

Photographs by John Bailey
Diagrams by Chris Turnbull

Typeset by Inforum, Portsmouth
Printed in Great Britain by Cromwell Press, Trowbridge, Wilts

Contents

Introduction

Perhaps we ought to state straight away that this is *not* a book attempting to sit easily on the crest of some trumped-up, imaginary perch revival. Our aim is simply to present a realistic picture of perch fishing as it really is in the late 1980s.

Perch are no longer with us in the numbers they once were, of course, but this does not make them unique. Indeed, a 2lb perch is a far easier proposition than a roach of the same weight in East Anglia today, but a true perch revival? Perhaps not that yet. Although the budding perch angler in the 1980s must not be afraid to blank, he must however *believe*, and if he does the blanks will perhaps be less frequent than he envisaged.

The dashing, fry-chasing perch.

Much of what we discuss centres on East Anglia and, to a lesser degree, Kent. We hope it will become clear, just as it became clear to us, that contemporary perch fishing is, happily, worthy of serious consideration again in certain areas in these regions. Furthermore, there is little doubt that this is also the case in many other parts of the country and not just those already mentioned. Isolated colonies of perch have been discovered and exploited in Dorset, the River Trent and some of the well-known Midlands 'circuit waters'. It is our belief that perch currently inhabit several different waters which are simply not being fished. Surely Norfolk is not unique in this respect.

A surprising amount of totally unspoilt perch fishing can be enjoyed by any angler who has the inclination to pioneer a little. Should you undertake such a quest, a degree of work, some research with a dash of courage and the all-important confidence will be necessary. Many anglers today already possess these qualities in their pursuit of other species, but they can now turn their attentions to perch, and be successful. We have not been carried away in thinking that most waters now contain perch again. They clearly do not, but certainly in a reasonable percentage of them perch are truly thriving.

Dawn sunshine thaws the ice at Greenmire Pool.

Introduction

To all those anglers who can recall catching those dashing, fry-chasing perch that were such a common feature all those years ago we say reconsider, and with effort, you can catch them again. For the younger generation of anglers who have sadly missed out up till now we say, go perch fishing and open up one of angling's finest treasures, something that you perhaps never thought you would experience.

As we pen this introduction on the last day of the 1987–88 season upon the banks of a perch water, the dusk is gaining on us fast. Of course the evil bacteria *Aeromonas* is still with us – it continues to cut down prime perch every year – but gazing out across the water here, it is a comforting thought that a 3lb perch can be as little as four years old in the right environment, so it pays to keep up to date with *every* water in your vicinity.

Now the darkness is almost total and it's time for the rods to be packed for the last time this season. As this final sentence is written, we both hope that the following pages will serve as some kind of spur to do something similar on your own waters. Should this be so then we will feel all the pleasure you will undoubtedly enjoy.

JB & RM
14th March 1988

Far Away and Long Ago

by Roger Miller

This memory, of a summer day far away and long ago, recalls an anti-cyclone so scorching it burned not unlike noon in Marrakesh. The fruit field seemed to simmer in a cauldron of stifled air, scented by the sweetness of its crop. The shade was heaven-sent when I found it in the spinney of beech and stunted oaks, far away from the sticky fields of blackcurrant that had stained my clothes during the morning of labour.

Lunch hour had at last arrived. 'It might as well rain until September' sang the definitive singer-songwriter of those lost and gone-forever days. The yoghurt had soured, the chocolate had melted and the cheese had curled at the edges. Lunch was out! I would doze by the river under the willow herb, perhaps even bathe in the coolness of a lazy summer flow.

She was restless and needed a shady hollow. The river did not afford many now that the working party had cut back the hanging bushes to give the dry fly purists an easier life. Downstream she meandered, not realising that the reappearance of cover coincided with the sign of the lowest end of the exclusive trout syndicate stretch, known locally as the 'Major's Mile'.

She eased past the common and under the road bridge. Two very young anglers jigging worms from above saw her ephemeral shadow appear and ran to the adult whose indifference to their 'fantasy' caused him to miss out on something extra special. On to the fruit farm stretch and on further to deeper water she went, after the rough turbulence of the common. More oxygen, dace shoals, clouds of minnows, roach too – perfect! The abstraction intake to the irrigation pond gave even deeper and slacker water still. She never noticed the dozing schoolboy beneath the willow-herb as her instincts told her that she had found the ideal residence for the duration of the drought . . . till it rained in September perhaps.

I had never seen a specimen so huge – indeed I have witnessed only one such specimen since. I had washed in the pond and then wandered across to the river and peered into the intake pool that had been created by years of water pumping on to the fields so bottles of Ribena could be made available for you and me. Now I had read Walker, adored Sheringham and been inspired by Venables, but nothing had prepared me for what I suddenly saw

I'd never seen a specimen so huge.

through the silver haze of the surface glare: a broad, thickset pig of a fish with fins the colour of raging fire and flanks striped with charcoal grey. Her aristocratic air intoxicated me on that summer day of sweet, wind-carried aromas and sunshine that lifted the skin off a bare back in a single afternoon of sweaty labour.

My formative plan appeared to be the correct one as the fish suddenly skittered through a cloud of minnows that had wafted just close enough to tempt her in her newly-adopted lair. My heart raced; my breath, fast and uneven, was the only sound I could hear as I cycled homeward. My mind was a kaleidoscopic mass of image and fantasy – the perch became an obsession. Venables was re-read, Walker revised, but Sheringham's *Coarse Fishing* confirmed the plan as I arranged to take some minnows with me to work the next day.

Dawn was taking forever to break through a mist dripping on and clinging to the valley like steam in a sauna. The perch was alert! This was the time of day she lived for, the hour in which she could tear through the minnows, the dace and the odd small roach in order to satisfy her hunger.

This then was the first dawn in her new domain, but the perch was soon

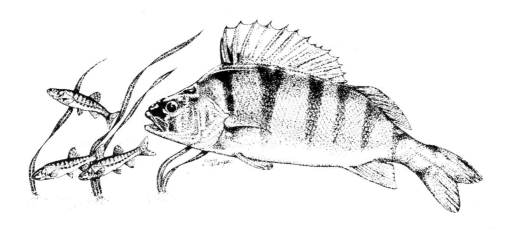

'. . . *as the fish suddenly skittered through a cloud of minnows . . .*' (*Drawing by Stephen Harper.*)

competing with the two pike that slunk in from nowhere for the rich pickings of the intake pool. The sun, till then obscured and invisible, suddenly proved too much for the chilling mist and burst through, flooding the scene and prompting the predator to nestle down ready for a day of lethargy. But she was still hungry – the pike had made hunting difficult for her, and the shoals of prey were split up. A fruit picker, unusually and uncharacteristically early, had just darted through the rows of blackcurrant bushes towards the pool. He did not appear to be carrying anything that resembled the tools of the fruit picker's trade.

It was of course a reckless thing to do, but boyish enthusiasm always overcomes petty rules and regulations. It was therefore with a certain amount of disdain that I tackled up beneath the 'Private. No Fishing' notice placed, of course, directly adjacent to the pool. Two fat little minnows, perfect as bait, were guardedly lowered down into the eddied flow. I had seen a couple of jacks out in midstream and was naturally anxious to avoid attracting them with the livebaits. The old perch took an instant dislike to my presentation and kicked off a few yards downstream, still clearly visible.

I allowed the bob-float to take the wriggling minnows, poor wretches, down to where she lay. The light was playing tricks on the surface and as I lost sight of the float in the glare, I lost sight of the perch too. The sudden animation of the rod tip left me in no doubt that she was on. Never had I known such power, such verve and cunning.

A diesel engine suddenly barked out from over by the farmyard; the foreman's dreaded tractor was on its way down to the fields – and here I was, fishing the pool, playing a fish, without permission. Surely I would be shot.

The perch, for the first time in her life was well and truly hooked. A size six was firmly embedded in the corner of her bony jaw and the 6lb line cushioned every run she made. The prey fish scattered in panic, perhaps believing the perch was feeding. The jack pike, like vultures, eyed the spectacle with keen interest. Up, down and across the flow she bored. She was being played very hard indeed, by a poacher who was intent on bagging his prey before the approach of the keeper!

With little ceremony, a single scoop of a tiny landing net extracted her from a watery, comfortable world to a dry, suffocating one. She sensed the burning hands of an angler upon her flanks, the gentle tug as the hook was slipped from its hold and if she could have understood the sensations along her lateral line she would have heard, 'What the hell are you up to young man?' from the foreman.

'She's a monster, isn't she?' said the fruit picker.

'Who the hell do you think you are?'

'A pretty damn good perch fisherman, actually. Got any scales?'

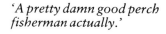

'A pretty damn good perch fisherman actually.'

'You cheeky young bugger! Clear off and hand over that fish.'

The perch, in no position to be grateful to her tormentor, found herself that instant back in a more familiar environment. The fierce discussion going on above the fast subsiding ripples was left behind as the old perch darted off, never to return to the pool again.

Boy, was I in trouble! Caught poaching the intake pool, returning the perch contrary to the wishes of the detestable old foreman and having the cheek to argue about it afterwards. The size of the fish had astounded me and it was not until years afterwards that I actually wondered about the weight. Then it was simply the achievement that meant so much; my standards were my own and nobody else's mattered.

Still the tractor driving foreman went on, the fierce discussion going on above the fast subsiding ripples was left behind as I too darted off, never to return to the pool again.

John Bailey

Nostalgia is easy and pleasant perchy memories can be snatched from almost any season out of thirty. The point is though, do they serve any real purpose? Are they any more than whimsical flights of fantasy that no reader should be expected to follow? To me, such recollections can reveal how a knowledge of the species is built up, experience by experience until it becomes useful if not complete. And just perhaps, tales from the past can fuel the imagination, fire the desires that motivate us all to get out there when the weather is foul and the fireside is fine. The memory of my first perch has stayed with me thirty odd years to this day. A hot August day. An old mill dam in Lancashire, choked with Canadian pond weed every acre of its dazzlingly clear water. My worm dangled a boy's rod-length from the bank in the merest clearing of the underwater forest. It was pink, I remember, and it wriggled and it enticed out a 3oz perch. (I know the weight well, for it was caught and killed, alas, and taken home and weighed.)

That little perch approached with caution, fearful lest it should be spotted and the worm escape. It used every scrap of cover until it was prepared to mount its final dash. Then its fins quivered, erect and bristling and it was nose to nose with the worm in a second. There, it huffed and it puffed and, with gills aflare, it gobbled the worm down. And my, didn't it fight till even my old cane poker rod gave a creak or two? And haven't I been in love with them ever since?

15

*A deep winter perch tempted in the last seconds of daylight
from Greenmire Pool.*

At the same dam, I hooked my first big perch. It happened like this, and
there is a moral here somewhere. You see, for once, I was allowed to fish
late. Again it was summer, 1959 I recollect, and the gas lamps on the
cobbled lane alongside flickered into life and hissed to the rats that night
was falling. The rats scampered here and there and everywhere along the
waterside masonry and with fearful determination I gripped the corks and
stared, blinkered, at my favourite red perch cork float.

Daylight faded right away until the red was mere shadow and I struck on
the slightest suspicion of its disappearance. What a fish I found myself into!
There was no throwing this one over my shoulder in the single flowing
movement I had perfected! My perch now bullied and bored and strained
the rod so greatly I had to follow him around the wall. Then he dived. The
rod followed him with its tip to the surface. The line stretched to its limit but
the anti-reverse was locked on and the clutch was screwed down tight with
my pliers so it was farewell to my first pounder.

Still, I went back to the reel manual and learned that a clutch could slip

16

Dusk – always a fine time for perch.

and why and I realised that dusk was a magic hour if only I could exploit it better.

To do that I really had to wait until I was adult and all through the 1973–74 season I fished Norfolk's Bayfield Lake hard. There as the autumnal and winter afternoons faded amongst the woods around, the good perch would always begin to move. Nearly fifty 'twos' came my way those quiet evenings and then, on Christmas Eve, along came my monster. The story has already been told in my first book, *In Visible Waters* and I won't repeat it here except to say that the fish did fight for over five minutes, I did see it at the net, and it would have weighed over four pounds. We never met again, but since, when perch fishing, I've never packed up before darkness has fallen.

After the death of the Bayfield perch, Worthing gravel pit taught me more about light values and then the pre-dawn and first light periods ruled supreme. There on still, steely days I saw perch hunt as never before or since. Their strategy was not unusual: they typically herded prey fish lightly into bays or against the bank. It was their attack that was so sensational.

Most perch strike from deep upwards to the surface and what one sees are small fish scattering and then a splash or a boil as the hunter turns. These Worthing perch, however, approached with their dorsals in the

'Those still, steely days.'

surface layers, and frequently these would show sail-like as the fish picked up speed and smacked into the food shoals fleeing before them. Perch fishing could not be more exciting. To see these big predators arrow in for the kill was savage, majestic, awe-inspiring, and when the bow wave was headed for your own float excitement knew no bounds.

I've caught perch from the Tweed in the North, to the Stour in the South. I'd love to hunt them in Holland and I've heard of genuine twenty-pounders in Australia – one day perhaps I'll get at them as well. And then, keeping it in the family, there are even Nile perch. With one of those at 150lb under my belt, I'd feel I'd come a long way from that Lancashire mill dam!

The Perch Disease

by Vic Bellars

Vic is an all-round angler of immeasurable experience with a special affinity for perch. A highly respected and popular member of the angling community, he is renowned for his tackle innovations but has undertaken considerable research into the perch disease *Aeromonas* and has kindly allowed us the chance to publish his unique insight and depth of knowledge in this field. A retired school teacher who now happily resides within 50 yards of the River Waveney, Vic spends much of his time developing the Marvic Fishing Tackle Company with friend and partner Martyn Page. Vic has an enviable perch record with individual specimens to over 4lb.

Once, not all that long ago, perch were a prolific species. Anglers, whenever they used worms or maggots as bait, were sure to include a sprinkling of perch among a catch of roach or bream. Hordes of seemingly ever-hungry perch could be found in rivers, lakes, reservoirs, canals and even tiny ponds. Small perch are bold-biting and suicidal, hence their popularity with small boys. A perch's penchant for a dangled worm turned many a boy into a lifelong angler. Such simple pleasure is denied small boys today, but there are signs that the species is slowly recovering from a devastating epidemic and beginning to recolonise its ancient haunts.

As an example of the abundance and fecundity of perch, during the last war there were so many in Lake Windermere that they were netted and sold commercially to bolster the nation's food supply. The fish were canned in oil and marketed as 'perchines'. There was no perch shortage in those days! Then, quite suddenly in the 1963–64 season, some of the Windermere perch began to die. The mortality percentage was low and the event passed unnoticed by all except the scientists at the Freshwater Biological Association Laboratory located at Ambleside.

Two years later an unprecedented disaster struck the species. Perch in the Metropolitan Water Company Reservoirs in outer London were virtually wiped out, the first water to become affected being the Queen Mary Reservoir. Hardly had the scale of the epidemic become recognised when *two tons* of dead perch were collected from Abberton Reservoir in Essex. It

Tomorrow's specimen – no disease here.

was not known how many tons of dead perch littered the bed of the reservoir, unrecovered. In 1968 the reservoirs of Pitsford, Ravensthorpe, Hollowell and Grafham were similarly affected. It was now apparent to scientists and anglers alike that a massive and unprecedented epidemic of unknown origin was spreading like wildfire from water to water. Other species cohabiting with the perch appeared unaffected but by 1976 when the epidemic reached its peak the roach in Shustoke Reservoir in Warwickshire and Blithfield in Shropshire also succumbed.

That same year the mortality rate among sexually mature spawning perch in Windermere was massive. It was estimated that 98 per cent of the total stock perished; at least one million adult fish. By 1977 perch waters nationwide, including rivers, were hit by what was called Perch Disease (PD). It was an unmitigated disaster of immense magnitude and anglers, their fathers and their grandfathers could not recall perch dying *en masse* before. Little was known or understood of the cause – both anglers and scientists were baffled and thought that some hitherto unknown disease never before encountered was responsible. For a time this theory held sway, but more recently evidence has come to light that heavy perch mortalities *had* occurred before. The disease had been noted in the 1870s. PD had killed perch before, not only in England but abroad as well.

Vic Bellars.

The Perch Disease

Marianne Plehn, a fishery biologist specialising in fish diseases and parasitic infestation, investigated mass mortalities of perch which occurred in Lake Geneva in Switzerland in 1924. These fish exhibited ulcerating lesions similar to those found on perch suffering from PD during the much later British epidemic. Even in North America the yellow perch, *Perca flavescens*, now believed by some to be the same species as the European perch, *Perca fluviatilis*, were affected. Isolated cases were investigated by Ross, Bailey (no relation), and Heaton in 1960. Their investigation concluded the cause to be bacterial, but no real causative agent was identified.

Because I like catching large perch and I knew so little about the disease that had struck in East Anglian waters where I fished, I sent a questionnaire to fishery scientists in every country in the Northern Hemisphere where perch are an indigenous species. I also contacted fishery biologists in Australasia where European perch had been introduced.

I asked among many others, two specific questions. Had they encountered mass perch mortalities, and if so, had they investigated the cause? Also, had they found any incidence of caudal (tail) fin degeneration? Not one failed to reply and those scientists in eastern bloc countries including the USSR answered promptly and as fully as they could. If perchance one of them ever reads this chapter I wish them to know how grateful I am for taking so much trouble to answer a simple angler's questions in layman's terms.

After collating the answers I was surprised to learn that PD was unheard of except in France. In 1979 there was a heavy perch mortality in Lake Annecy, an alpine water. The dead and dying perch suffered from epidermal lesions such as had appeared on British fish similarly affected. During the same year further mortalities occurred in lakes at Auvergne, and in both cases investigation failed to discover the cause of the disease. So it appears that we in Britain have suffered the worst and most widespread outbreak of PD ever recorded.

Tail fin degeneration appears to be another phenomenon peculiar to this country as far as perch are concerned. Bacterial degeneration of the caudal fin is known of abroad, but occurs mainly in zander and occasionally in carp. Tailless rainbow trout are far from uncommon.

The fact that there seems to have been so little incidence of PD abroad does not imply that it has not occurred. Isolated cases in sparsely populated areas could well have gone unrecorded. Unless perch are an important food resource little heed is likely to be taken of anglers' complaints, so perch dying here and there would be of scant interest, except of course to an angler.

Perch with caudals intact.

In Hungary during the last decade, perch have virtually disappeared from many waters. Two lakes, Balaton (60,000 hectares) and Velence, (2,200 hectares), both once prolific perch waters, have few perch remaining. Rivers too, are suffering from diminishing stocks. I can obtain no record of diseased fish but when millions of perch suddenly disappear it is unlikely to be from natural causes. The Hungarians are of the opinion that the lake perch were ousted by an increasing eel population feeding on the perch spawn, and the dearth of perch in rivers is due to simple pollution. These explanations might be valid, but somehow it seems an improbable theory particularly with regard to huge lakes. So there is some evidence, albeit sparse, that PD has occurred elsewhere other than in France. What remains to be discovered is whether the causative agent of PD is the same in all countries.

Freshwater fish can be affected by many different bacteria and perch are no exception. For instance, any damage to the protective mucous or epidermis such as is caused by the bite of *Argulus foliaceus*, the fish louse, can lead to fungal infection. It seems unlikely that *Argulus* infestation is a contributory factor in the spread of PD. Certainly *Argulus* could not have inflicted the lesions symptomatic of the disease.

Anglers need to be able to recognise a perch infected with PD; they are also keen to know the cause, and whether our perch stocks will ever fully recover. The question of what causes PD cannot be answered, for research has to date been inconclusive, although certain factors do indicate a possible cause.

If PD had affected North Sea cod, or Atlantic salmon, who have had troubles enough with UDN because these species are important commercial food resources, a full research programme into the cause would undoubtedly have been instigated. But perch are unimportant, except to the angler, so what little research has been done on the problem was the result of interest on the part of fishery scientists themselves.

The Freshwater Biological Association has been conducting studies on the perch of Windermere for many years and as chance would have it there was a massive outbreak of PD in that very lake.

Two important papers, both published in 1979, studied an epizootic of perch. Since then little further work of PD has been undertaken. The findings in both papers, which make fascinating reading, even for a layman, are inconclusive as to the actual cause of PD.

PD is a perch-specific disease and is characterised by the appearance of lesions on the body and fins. Usually these lesions are more common on the posterior flanks and at the base of the caudal fin. These ulcers can, as the disease progresses, result in deep necrosis which exposes the musculature and skeletal structures. These lesions have associated oedema and haemorrhage. In other words the fish exhibit deep, raw, bleeding wounds, the edges of which are colonised by bacterial and fungal infection as a secondary condition. At the same time internal organs are affected, the gills and viscera (intestines) become pale and anaemic. It is almost certain that any perch that is infected becomes moribund, and will subsequently die. The carcass may be washed ashore, but most sink to the bottom.

In Windermere, no infected fish recovered as far as is known, but during a health check on perch at Grafham Reservoir in 1975, seven years after the initial outbreak of PD, a few fish showed evidence of healed ulcers. The regenerated tissue could have repaired damage caused not by PD but by some other misfortune, so while the possibility remains that some fish can survive PD, it has yet to be proved.

With regard to recolonisation we are on surer ground, for there is ample evidence that our perch stocks are slowly increasing. Localised epidemics of PD occur from time to time but the trend continues.

The fungal infections which are a secondary condition of PD are more prevalent in winter. In Windermere 58 per cent of samples of diseased perch showed signs of fungal infection but only 14 per cent of samples showed the same signs in summer. It was also interesting that compound fungal infections were noted, with up to a maximum of four different organisms growing on a single epidermal lesion. During the investigations a wide variety of aquatic fungi were isolated from infected fish, the most common being *Aeromonas hydrophila*, particularly prevalent in summer. As this organism was also isolated on healthy fish, and was absent from some infected fish, it was not considered a likely cause of PD. As myxobacteria were only found infrequently it was also decided that PD was unlikely to be a Columnaris-type disease, that had in the past reached epidemic proportions among roach. However, the furunculosis *Aeromonas Salmonicida* sub-species *achriomogenes* was identified on a number of diseased Windermere perch, normally associated with salmon, which was of great interest.

In April 1978 it was reported that perch in a Cheshire lake were showing signs of PD. This gave the author of the papers, Mr Bucke, the chance to collect samples for keeping in tanks at the Fish Diseases Laboratory at Weymouth. There was no sign of heavy fish mortality, but the samples showed evidence of ulcerative skin lesions.

Three fish were killed at once for histology, bacteriology and virology tests. The post mortems revealed PD-like lesions on the caudal flanks with skeletal muscle exposed. The visceral organs were pale, but no other abnormalities were noted. Bacteriological tests on heart, blood and kidneys proved negative. But as before, bacteria isolated from the lesions was confirmed as *Aeromonas hydrophila*. Virological tests were also negative.

Experiments with the remaining live fish, some diseased and others healthy, obtained from a water with no history of PD whatever were as follows:

1. *Transmission experiment*   This was to discover if the disease could be transmitted by cohabitation.
2. *Stress experiment*   Diseased fish were injected with an immuno-suppressant to exacerbate the disease state so they would release more micro-organisms. Healthy fish, as controls, were placed in the same tank and also immuno-suppressed.
3. *Virulence Experiment*   Healthy perch were infected with a bacterial isolate obtained from previous tests on diseased fish.
4. *Transmission experiment using the same bacterial isolate as in the virulence experiment*

(a) Three fish were inoculated with the cultivated organism and kept with three non-infected perch. These fish were kept in flowing water.

(b) Six perch were held in a tank of still water to which the test culture was introduced. After two hours a continuous water flow was applied.

(c) Six perch were fed worms which had been infected with the culture.

(d) Six perch were anaesthetised and a small scarified abrasion made on the right flank. The culture was introduced to the water and continued as in experiment (b). In all cases the test fish were diseased, the control fish healthy.

THE RESULTS

I have attempted to simplify these so as to be of interest to the non-scientific angler.

Experiment 1

Within three days the test fish developed a fungal infection at the side of the existing ulcers. The three healthy control perch became covered in mucous lumps after 15 days. Apart from this all fish remained alive until day 32 when a control fish died, followed on day 36 by another.

The post mortems did not reveal any PD-like lesions, just exuding mucous on skin and gills. The test fish showed similar symptoms and even signs of the ulcerative lesions' separation. Some fungal growth occurred on the outer layers of skin. This fish also exhibited histological changes similar to the controls. No bacteria were isolated and the control fish did not show significant abnormalities.

Experiment 2

The two test fish developed a pale raised oedematous area in the dorsal region. One test fish died two days after injection, another was killed on day 6. No signs of PD symptoms were noted. The post mortem revealed that the oedematous area contained purulent fluid. Samples of this were taken for bacteriological examination. Internally, the visceral cavity also contained the same fluid; the visceral organs and gills appeared pale. The raised lesion, unlike PD, had not ruptured the skin.

The examination detected micro-colonies of bacteria inside muscle bundles and a marked leucocyte (white blood cells) infiltration of the lesion.

Some mitosis was found. Mitosis is a process whereby the nucleus of a cell undergoing cell division (that is, splitting to form new cells) divides to form two new nuclei each with the same number of chromosomes as the original nucleus. Other changes of little interest to the angler occurred; no bacteria were found in any internal organ. However, tests revealed 1mm colonies after 14 days' culture. A later test for *Aeromonas Salmonicida* proved negative, but when the isolates were further sub-cultured on to a fresh medium, a positive reaction for *Aeromonas Salmonicida* was obtained. The control fish exhibited no abnormalities, so even fish made artificially more susceptible to disease remained free from infection, although closely confined with fish showing symptoms of PD.

Experiment 3

There was a noticeable reaction at the base of the injection site which became evident 6–12 days after treatment. Four fish died, but *Aeromonas Salmonicida* was found at the injection site of two moribund fish. Lesions had developed downwards from the injection site with marked muscle degeneration.

Experiment 4

(a) All inoculated fish died within three weeks. The control fish remained healthy. Bacteriological tests were negative.
(b) These fish did not seem to develop any symptoms of PD from the infected water. Two fish died at 20 and 28 days respectively, for no apparent reason.
(c) The infected worms did not appear to affect the perch in any way.
(d) *Scarification* One fish died after ten days, but did not show any signs of the disease. The remainder remained healthy. After thirty days the fish were killed and *Aeromonas Salmonicida* was isolated from the heart blood of two perch only.

Above is a simplified account of the work undertaken to determine the causative agent of PD. I have excluded all the work on secondary fungal infections, but many kinds of differing organisms were detected. No particular one or group of these can be held responsible for PD and they should be considered as secondary infections. What is of particular interest is that the cohabitation of healthy fish in close proximity to diseased ones affected them little. So we can conclude, perhaps with caution, that PD is not virulently contagious. This, however, is difficult to equate with the fact that PD killed one million perch in Windermere so quickly.

Certainly, *Aeromonas Salmonicida* rears its ugly head at times, and is the causative agent of furuneutosis, but it did prove difficult to detect among the other faster growing organisms such as *Aeromonas hydrophila* which colonise an ulcerative site. So we still do not actually know what causes PD. The best candidate must be *Aeromonas Salmonicida*, but it is equally possible that more than one causative agent could be responsible. What is certain is that the disease is caused by a bacteria rather than a virus. Just recently, I have heard that a viral disease has affected perch in Australasia, but I have no details.

One common factor is that PD often develops after spawning, a high stress period, and an outbreak is more likely at such a time. Windermere had an infection commencing in January which continued to past spawning. It is possible that whatever organism causes PD, it could be triggered off by stress in the fish themselves. This could be as a result of debilitation from spawning exertion or a combination of environmental factors at which we

We hope to God that fish like this survive.

can only guess. We ourselves if overworked or run-down are more likely to catch a cold or flu, so perhaps stressed and tired perch are more susceptible to PD. Wild, unscientific surmises often contain a grain of truth, so who knows?

Myxomatosis did not eradicate our wild rabbits – a partially resistant strain has evolved in a remarkably short time. Young rabbits gain immunity to the disease from their mother's milk, which protects them for at least the first year of their lives. So too, a PD-resistant strain of perch may well develop, or is perhaps already beginning to do so. As an angler I look forward to that!

I would have liked to have concluded with a definitive statement of the cause of PD. Perhaps *Aeromonas Salmonicida* is responsible, but I fear it will be many a long day before we know for certain. So PD remains an enigma so far. It has been with us for longer than is generally realised. There will be epidemics to come – as I write, the disease has reappeared in Kent and at any time East Anglia's mini perch revival could be decimated. But some will survive and recolonisation will take place again, as before – that at least is a comforting thought.

COMMENT

It is a sad indictment of the water authorities that no further action has been taken in trying to further the good work described above. A perch fishing colleague in Kent is currently using his professional expertise during his spare time in an effort to learn more about this most perplexing subject. We look forward to his findings with considerable interest. In the meantime all we can do is hold our breath and pray that the time bomb just waiting to explode in our perch waters again does not do so before a lot more facts are known.

JB & RM

The Motivation for the Project Ahead

by Roger Miller

During the summer of 1984 the first inkling of the new generation of perch in Norfolk came our way after an all-night session for big bream on a Wensum Valley gravel pit. At dawn we noticed another angler making his way round the underfished, overgrown wilderness of the opposite bank before settling in the end swim that nobody ever bream-fished. He appeared, however, to be fishing for pike, as well he might, for every dawn would see them striking at the multitude of tiny roach that also swam in the large, wind-swept pit.

Little notice was taken of his playing of the occasional jack throughout that morning, but through sheer boredom at the lack of bream activity I decided to pay him a quick visit.

'Heard about the bream I see,' was his greeting.

'Yes, none today though. You've had some jacks I see,' I replied.

'Typical of the place of course.'

I should have put two and two together there and then, for I noticed the sack in the water obviously containing a fish. Nobody sacks jack pike!

After lunch, when even his sport died, we left, and never saw him there again. Upon returning to the pit a couple of days later a boy told us about the 'pike angler' who, amongst scores of jack pike, had taken a small number of specimen perch. We were flabbergasted. We recalled all those strikes on the prey fish we had seen, how a red-topped bait marker would be taken on the retrieve amidst a flurry of spray and finally the 'pike angler' and his sacked fish. The water had also enjoyed a reputation for its big perch before disease wiped them all out, years previously. The place had then been forgotten about, almost! We decided against this very tempting diversion and continued with other pre-determined projects and gradually the perch at 'The Bream Pit' were forgotten.

The following spring saw us in a working party netting some stew ponds for their jumbo rainbows. These were to be transferred to the main lake that

The new, unblemished generation of perch.

adjoined the stew ponds by a series of interconnecting, underground pipes. Throughout our labours that day, literally hundreds of baby perch were flushed out of the pipes and caught in the nets. This had never happened before except with roach, and although none were over three inches in length, it was clear that there had to be a healthy stock of adult fish in the main trout lake – indeed odd ones had even been caught. Their fry were probably inhabiting the pipes for sanctuary. There was never going to be any chance of perch fishing the actual lake so neither of us gave it a second thought.

Perhaps the most startling discovery for us was made during the late summer/early autumn period of that year during a hard but wonderful roach quest on the Wensum. That particular campaign was not only memorable for its roach but for our discovery of perch. On one particular evening that was painted with a million shades of red at dusk we saw fry topping everywhere in the stretch. Yes, I am still talking about the poor, dying, old Wensum, but this was not the only surprise, for suddenly the bend that was our swim erupted with perch. Although only about a pound in weight, they were chasing, harrying and striking everywhere. Had we not

31

Perch like this one quickened us to the quest.

have seen this with our own eyes we would not have believed it from anyone! To compound this phenomenon, exactly the same thing occurred two miles downstream two evenings later. Then, the possibility of a specimen perch was realised as an obvious two-pounder was spotted swimming upstream four miles from the second sighting.

At this stage, not surprisingly, we began to think more and more about the various perch we had seen. Although never believing that a perch revival was under way it was quite evident that colonies of perch were indeed beginning to flourish again in certain waters, waters that had a pedigree of big perch in the dim and distant past. Enticingly, the two-pound Wensum perch was seen in the very area where an authenticated four-pounder had been taken in the 1960s.

The whole issue quickly began to snowball with two more perch waters suddenly springing up. John Bailey witnessed a holiday angler take several good perch from a totally unexpected venue only to hear him say that his best that week was an astonishing 3lb 4oz. With the perch feeding actively and visibly it was clear that the biggest perch in the water had yet to be taken. Within a few days another holiday angler pointed the way with a 3lb 6oz perch caught on lobworm whilst tench fishing. He had simply seen the big perch striking at roach, decided to cast his float fished lobworm to it and the perch duly obliged him!

Unfortunately both these waters were denied us after even the politest enquiries, but we felt sure that other waters could perhaps emulate them if we could only find them. We sat down one evening at our local with pen and paper and listed every snippet of information we had gleaned over the preceding two years or so. After a lot of memory work we were genuinely astonished at the wealth of hints, rumours and suspicions we had collected in that time. And yet, true feedback from other anglers in the accepted sense was virtually non-existent. It was always what had occurred during a tench, bream or pike session. Even if the presence of perch was always indicated, their downfall in each particular case would still have to be started from scratch.

Retrospective feedback, however, was very valuable: all the information gleaned from the 'good old days' merchants who were perfectly willing to divulge any amount of information about the glory days of Norfolk perching before the disease. They would gladly reveal the previously clandestine locations of the very best pits, lakes and river stretches. We were offered the most effective methods of the time, all of which made the project even more fascinating. Little did any of them realise that all the knowledge now thought worthless was in fact extremely pertinent to the autumn of 1987. Alas, with time as tight as it is these days we could only try some of the best waters at our disposal for the purposes of this book.

The motivation for the quest was of course fuelled by the prospect of simply catching good perch again, but in truth it went a little deeper than that. I would say that I was unfortunate in that entry into specimen hunting, as it was then called, coincided with the virtual wipe-out of East Anglia's perch stocks. In the autumn of 1987 I felt that even with my reasonable grasp of contemporary angling methods and tackle developments, there was still very little of it which was applicable to perch fishing. A wonderful new prospect had presented itself with only boyhood memories to go on. I felt as if I was starting out again with no prejudices, preconceived ideas or trends to sway me. Both of us were very conscious of the fact that perhaps somewhere, just somewhere, a truly massive fish was there for the taking – a true leviathan amongst perch was a distinct possibility!

With the whole of the Norfolk angling community either chasing broad-

The strike!

Gently does it!

The display.

land pike, Ardleigh perch or simply metaphorical rainbows here and there, we both knew that a whole county awaited perch exploration without hindrance or rival. The chance to break new ground, to be totally detached from the local scene and to discover some nice perch seemed heaven-sent.

Further motivation was sparked by a delve into our angling libraries in order to study, review and refresh ourselves on the long-forgotten ways and wiles of the perch fisher. Inspiration is a by-product of such research but hours of reading revealed that perhaps Richard Walker and to a lesser extent Ken Taylor were the only two anglers who had understood perch thoroughly and fished for them with real knowledge. We realised that unique waters require a unique approach, of which more later.

This, then, is the basis from which we embarked on the venture. We jumped into the void, totally cold, in pursuit of a species phoenix-like in its reappearance. Our appetites for the coming months were insatiable. Our motivation was total, the blank pages of fate lay before us, pregnant with possibilities . . .

Small Waters

by John Bailey

Experiences have shown over several years that no water is too small to be a potential big perch venue. Quite truthfully, I have a friend whose garden pond produced a 4lb fish. The key to this perch's success was a superabundance of tiny goldfish. The perch did not need space as much as a real supply of food. The lack of space even worked for it, making the little fishes' getaway even more difficult. I suppose, in a ridiculous world, we could all create our own large perch 'fisheries' using this example. More realistically, Jim's pond ought to teach us never to overlook the smallest puddle of water – a lesson Roger and I nearly, disastrously, failed to recognise one dour dawn at Ha Ha pit. It happened like this . . .

HA HA PIT

One October – October 1986 to be precise – I was fishing for perch on a lake where it happened perch did not exist. Sport, I will admit, was slow and I was glad indeed to see old Bill come around for a chat as the light fell on a gloomy, dull day. How he laughed when he knew what I was about.

'You're on the wrong lake, mate,' he said. 'There was a big perch taken hereabouts, but not from here. You're a couple of miles out. You want to be the other side of the woodland here.' And so I was told about Ha Ha.

I packed immediately and drove to find it in the darkening countryside. It was no easy task, as it proved to be hidden by a ring of bramble, blackthorn and alder. It was tiny, no more than half an acre, and as I crouched by it in the last of the light I was sure this tip too would prove to be false. Then a little shiver of magic entered the evening: a shower of small fish came out at my feet and a flat of water showed where some predator had turned beneath them.

Plans were at once made. Permission was sought and most grudgingly given for a single visit. And so it was, in November, in the half-light of a blustery day, Roger and I stood for the first time rodded up and ready to go at Ha Ha.

A pike from a 'Perch only' water causes Roger Miller some disappointment.

'John,' said Roger, 'You are joking? You must be. You can't have got me up at five for a puddle like this.' I too was taken aback to see just how small it seemed in the growing day and was on the defensive.

'We can give it till ten and then can always move on.' I suggested. We hardly knew then how our emotions would yo-yo in the following hour.

They raised after only ten minutes when a few yards out small fish scattered in a boil of leaden water. Two baits immediately entered the attack zone where Roger's gudgeon was taken. The fish fought deep and, if it had been a perch, it would have been massive. But a 6lb jack pike was not what either of us wanted to see. Our spirits collapsed.

We were on the verge of leaving Ha Ha after only an hour. I thought back to the strike I had witnessed that first evening visit and knew how easily a small pike could have been the culprit. And even when the float above my deadbait moved against the wind and also gently dipped beneath the ripple, I felt little excitement.

Line ran out and I hit it early. From the start I knew this too was a pike. Heavy and sullen it drove up the tiny pit but I hauled it back towards me. No jack would take me where it wanted the mood I was in. At the rod tip the pike rolled, about two feet down, in the crystal clear water. It was hump-backed. It had blazing red fins. I could see broad black bars on its flank and a dagger-pointed dorsal fin flaringly erect. I collapsed. The wind went out of my body, the bend out of my rod and my wits everywhere. Helpless. Hopeless. I jabbered at Roger again and again.

'Net it. Just net it. For God's sake net it.'

Half to his waist in water, Roger did just that and, carrying the beast ashore, the dead roach and my hook slid completely free from the great fish's mouth. Now, every angler knows moments of elation – I've been lucky to taste several – but this was pure champagne. We danced. We cheered. We hugged each other. After less than an hour Ha Ha was immortalised!

How much would it weigh? We looked at the fish lying in the micromesh. After 115 perch over 2lb to 2lb 15 oz, I knew this was the '3' I had badly wanted for over 12 years and had believed lost to me forever. How much over 3lb the perch would be I could not tell and did not really even care. But at 3lb 12oz, you can believe me that even singing broke out over that tiny water.

The next hours passed by in a golden glow. Two-pounders hit our nets. We had a spell of fish between 1 and 1½lb and as dusk fell the fry once more began to leap and once all four rods were in action to fish of 6oz, 1lb 10oz, 2lb 1oz, and 2lb 6oz. At the close of play that glorious day we

The head of a monster.

reckoned we had netted close on 40lb of perch. They were all splendid fish with their tiger striping, and I do not believe either of us has enjoyed a day's fishing more in careers that between us stretch for nigh on 60 years.

MILL POND

Not all small waters have proved as fertile as Ha Ha! Indeed quite why there are so many good perch there remains something of a mystery. We have gone back on occasions and always done well with new, uncaught fish gracing the net. Perch are very easily identifiable by shape, fins, stripes and colouring and we guess to have landed some 20 different big perch there. Affairs have proved to be quite different at Mill Pond and though the two pools appear to have much in common, fish stocks vary vastly.

Mill Pond is also small – no more than an acre – and was made relatively recently. It possesses all the qualities necessary for a big perch water: clear water, luxuriant weed growth and a good stock of small prey fish – in this case roach and rudd. There are also pretty decent carp and it was for these

Another two-pounder hits the net.

that John Nunn and I sat down on the banks, in my case for the first time, one August day in 1985.

I seem to remember that we fished particles on one rod and boilies on the other and that very little happened for a very long time indeed. A carp or two tented in the thick Canadian pond weed but that was about all to disturb our discussions on football, fishing and the state of the world. In the mid-afternoon, quite unexpectedly, John had a quick run on the boilie rod. He struck and we were both disappointed at what was obviously a small fish. John bundled it in towards the net where we were staggered to see, of course, a perch. The tutti-frutti boilie, or whatever, hung from its mouth and it looked at us very crossly.

It was a lovely fish, quite perfect in every detail but for a mole-type growth on its flank near the pectoral. It weighed exactly 2lb 6oz and excited us greatly. Here, it seemed, lay an undiscovered perch water, so we caught some tiny roach and fished them unsuccessfully until dusk when the carp once again called to us.

It was November when I finally returned for perch, with a pail of worms, a tin of maggots and a supply of small freshly-killed gudgeon. On and off

The Mill Pond perch in close-up.

through the day I took roach, a small carp and, very occasionally, small
perch of up to 8oz. My spirits never once dropped, though, and I felt I
would be rewarded. Once again, in the mid-afternoon, so I was. The battle
on lighter gear was exhilarating and the perch weighed 2lb 7oz. There was a
mole in its flank. It was John Nunn's fish, fit, fighting and an ounce
heavier.

Some weeks later, I visited Mill Pond with Chris Turnbull and once again
the pattern of small fish was repeated. The only bait to remain untouched
was a small livebait Chris had working some yards out on the fringe of the
near dead clump of Canadian. For hours it kicked there but suddenly grew
lively and its float went down with a plop. This too was a lovely perch. With
a distinctive mole! Now our most obliging friend weighed 2lb 10oz, and
more and more it seemed as if he were the one decent perch in the place.

But it was not quite so. Later that day I had a good bite on worm and
pulled into a superb fish. It thrashed in the clear water and for a moment
hung over the lip of the net. I saw him quite clearly before the hook-hold
came away. He must have weighed 3lb or more. I was devastated. And so
too was John Nunn a fortnight or so later when he also dropped a big perch.

It would seem that at least two big fish patrolled Mill Pond then, but the quantity was nothing like what we discovered at Ha Ha.

In 1987 it obviously occurred to us to re-check Mill Pond, on the progress of the two big fish, and on how the smaller fish were coming on. Three Sundays saw just two fish landed and neither would have made the pound. Perch were still there, obviously. There had been no wipe-out disease and yet they had not really come on and done well. It seems that at Mill Pond the species retains a toe-hold but little more. Something in the make up of the place stops it reaching the heights of Ha Ha – as yet, that is, for nothing can ever be said quite positively about small water perch.

The Farm Pond

by Roger Miller

At the moment of writing on this mid-winter's night, I imagine him trudging from some obscure estate with a couple of pheasants tucked under his arm. Or perhaps beach casting for cod in some old smuggler's cove on the north coast. Or even regaling the regulars of his local country pub with oft-told tales of rural life. He is one of the handful of countrymen now left whose angling successes through the years would astound the angling world if only they had the slightest interest in publicising them. Whenever they speak we listen, and on this occasion he told us of a farm pond, after appearing heron-like at dusk as we packed our gear prior to leaving the forest pool . . .

The farm pond he spoke of was both deep and snaggy, hidden, amazing-

The tiny farm pond.

44

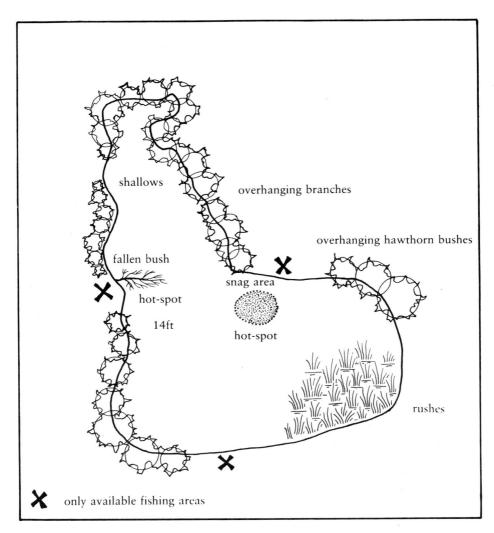

shallows

overhanging branches

overhanging hawthorn bushes

fallen bush

snag area

hot-spot

14ft

hot-spot

rushes

✖ only available fishing areas

The Farm Pond.

ly, at the top of a hill, and a three-pounder was rumoured to have been taken accidentally there the previous summer. Now I genuinely adore fishing somewhere on just a rumour and, as the summer began to flag, I called a respite from the lift-method and sweetcorn and decided to seek out this mysterious pond that was not even on the map. Through farmyards and bridleways I marched and, just as he had said, there it was on the crest of a hill. Two tennis courts in size, with a colour typically that of farm ponds, it lay before me. Perch seemed a remote possibility at that moment, but as it was 14 feet at its deepest point it appeared to have certain features common to big perch domains.

I squeezed between the unkempt bramble and hawthorn and as I cast a free-roving carp livebait to explore the pond 10 feet down I was suddenly taken with a sudden burst of inexplicable optimism. It was also the day that my personal change from glass to carbon was made and for the first time in my life I was actually proud of my tackle. Before I had a chance to set up and admire the second sleek, black rod the tiny pilot float audibly plopped as it disappeared from view. The 4lb line streamed from the spool and I recall thinking that if the three-pounder was only a rumour after all, I was still aching to see what unknown, confident predator had now taken the carp so unceremoniously.

The bale arm snapped over, the line was checked and the rod was christened as it jagged in response to the head-shaking perch deep down in the coloured mire of the pond. I pumped it up from the bottom, but gave line as it bore back to the depths again where the characteristic head-shaking continued. At last the stop knot appeared above the surface and I knew I had her, but suddenly and unaccountably the line began to grate. I could feel the perch struggling still but the contact felt dulled and indirect. Then I could reel no more – I was snagged! I hand-lined from above, from right and then left. I threaded the rod round the pond's perimeter foliage and hand-lined from the opposite direction but it was useless. The snag was immovable. Near to hysteria I hand-lined again but with the clichéd 'crack' the line parted and slowly fell limp and lost. With blind rage I gripped the carbon rod and with all my might smacked it against a tree. The second 'crack' of the afternoon resounded down the valley, closely followed by an equally loud curse. One minute I had been playing a big perch on a fine new rod and now in seconds the first was gone and the second was broken. An angler's life! I tied the rod up with string but remained biteless till dusk when I packed up the tackle in less than my usual good temper.

Within a couple of days I forced myself back with a mended rod and repaired mood, a bucket of carp and dace and a box of lobworms dug from the orchard the night before. The overhanging bushes I liked. I felt a float-legered lob fished tight to the branches would perhaps tempt a big perch out of its lair. I persisted to search the middle deeps with a free-roving carp livebait, and a float-legered freshly-killed deadbait lay appealingly in the reeded bay. Heavily punctured with a baiting needle I was convinced that the scent of the bait would permeate through the pond rapidly and more efficiently.

The 3lb eel from the farm pond.

The final session at the farm pond.

The rain was torrential, and as it began to dilute my belief in the place, the livebait began to bob around before shooting off in obvious indication of a positive take. A strike, a kick down below and it was lost. An immediate re-cast saw an identical repeat with the dace being cut virtually in half. The third time proved lucky as at last I found myself playing a fish from the depths of the farm pond.

I knew it was an eel as soon as I managed to prise it off the bottom. Five minutes later it was at last writhing in the net, weighing-in at a couple of ounces over 3lb – just the weight I was looking for if only I could get the species right! I thought briefly that it could have been the fish that had snagged me on my previous visit, but decided totally against that as the fight of a big perch and a big eel are so unmistakably different. I went on to suffer three consecutive blanks after the eel caper and on the fourth session I had almost decided to throw in the towel and change waters.

The afternoon was mild, gentle and soft and suddenly specimen crucians began to roll in numbers by a large, overhanging bramble bush. Not having caught a big crucian for 15 years I crept round and float-legered a lob's tail on a size 16, to a 2lb bottom with a two BB stick float for bite indication. Drawing the tackle as close as possible to the trailing branches I waited for the faintest indication of a bite so typical of crucians. Eventually tiny rings appeared round the $1/8$in of the float tip still above the surface.

Then, just as I was playing the 'crucian', a perch of 1lb broke surface and tried to spit the hook back at me. I do not know who was more surprised, me or the perch, but never have I been so pleased with such an unremarkable fish. John Bailey duly arrived, noticed the method and had a couple more before the crucians stopped rolling, which inexplicably stopped the perch feeding too. Unfortunately we never had a hint of a crucian. The ways of fish!

I decided that only two more sessions were to be devoted to the pond. The fine line tactics right up against the trailing branches picked up a couple more perch of the same size whilst every other method failed abysmally. On my 'final' session there I had convinced myself that the lost big perch was in fact an old eel after all but as I stood on the ridge taking photographs I suddenly discerned, through the lens, fleeing roach jumping for their lives in the reeded bay. I scampered down to the pond and saw a broad, spiny back bow-waving into the cornered shoal. I smiled wryly to myself as I realised that I had no livebait with me and the last of the worms had been used as loose feed beneath the hawthorn swim. I really admired the timing of that old perch that had at that instant changed the rumour into a hard, cold fact.

The conclusions that I drew from this fascinating little water were to enhance my respect for these most perplexing of fish. I had fished there on eight separate occasions. I had undoubtedly covered a lot of perch with either worms, livebaits or deadbaits on every trip, due to the smallness of the water and yet only fine line and almost match-style tactics had really worked for one of this country's most voracious fish, the perch. I felt pleased with myself in this respect as the perch will undoubtedly go undetected there for years, except by the old countryman in the know. I have not seen him to ask advice; perhaps he has none.

My eternal regret is of course the loss of the big perch on the first trip. I have returned many times since last seeing him when I had no bait but to no avail. The rod has since been repaired as good as new — unlike my spirits whenever I think of that abrupt grating and snagging of the line and the crack resounding through the mistiness of that grey, sleepy, north Norfolk Sunday afternoon.

The Sand-Pit

by Vic Bellars

Another angler told me about the pond. 'Someone's had a 2lb perch,' he said and then intimated that he intended to try and catch one for himself. Although the pond was only three miles from my home I had been unaware of its existence, and had I known I would only have been interested in the water as a possible source of livebaits. Interesting, I thought, but at that time just before the season opened, I had planned to catch some bream from a large lake. I could stroll down to the lake and be fishing within five minutes of leaving the house. Having baited up a swim every evening for two weeks I had no intention of altering my plans. Besides, I like to fish for perch in September and October, so if there really were any good perch in the pond I would find out soon enough. There was no hurry.

At that time I was teaching at a comprehensive school and was responsible for a group of older pupils, who were not renowned for their behaviour or academic ability. They were not unintelligent, just disillusioned, disinterested and downright idle. A number of these stalwarts were keen on fishing and adept at turning my carefully prepared lessons into angling seminars. They did not succeed often but now and again I purposely allowed myself to be side-tracked. We spent the odd lesson tying knots, whipping rings on rods and making up float and ledger rigs.

I remember one morning heading towards my classroom after morning break thinking, 'That lot again. They are going to do some work or I'll screw them into the ground.' As I reached the door I was surrounded by a group of lads all shouting at once. When quiet had been restored I elected the most vociferous to explain the reason for the hubbub. Very politely for him he announced excitedly, 'I've seen a bloody great perch, sir,' extending his arms to indicate its length. Any perch that size would have weighed about 15lb! Once the class had settled down I asked for the full story. It appeared he had been fishing the pond and had seen a very big perch swim lazily through his swim, so close he said that he swore he could make out every scale. The content of the lesson, 'The value of counterchange in graphic design', had of necessity to be changed to 'How to catch large perch'. I had no option. It was a good lesson, the boys wrote badly-spelt

51

copious notes, made drawings of rigs and behaved impeccably. And I began to think about the pond quite seriously!

During the summer holiday I had to visit the nearby town to get some line and oddments of tackle. On my way home I decided to look at the pond. After some searching I found it, close to a refuse tip and not far from a housing estate. There were two ponds separated by a pathway. It is difficult to estimate an area of water, but perhaps one pond barely exceeded an acre and the other seemed a little smaller.

It seemed every swim was occupied by three or four children and others were running about – it was mayhem. However, the largest pond seemed deep and very clear, and, as I later discovered, the bottom was solid with Canadian pondweed. One end was covered with potamogeton and there were other areas where this narrow-leaved water-lily grew profusely.

I chatted to some of the children and learned that they were trying to catch tench. Others were catching little rudd and two boys said that there were some large perch in the pond, but they didn't know if any had been caught. I drove home thinking that a well-weeded, deep, clear sandpit full of tiny rudd would suit perch very nicely.

One weekday evening in September I arrived at the pond with just one rod and a bait box full of large lobworms. One ancient angler was fishing the smaller, shallower pond so I could spend an hour or so checking depths and wandering around the other side without disturbing anyone. The pond was longer than it was wide and the far side was overgrown with willows and brambles. Just outside their trailing branches was a narrow fringe of potamogeton. It was impossible to carve a swim on this bank, so any perch could shelter there when the pond was disturbed by the hordes of young-sters in the summer.

Because a strong breeze was blowing down the length of the pond I felt that a float-suspended worm would be impossible to hold close to the lilies opposite. Legering was out of the question because of the heavy bottom weed, so I set up a scaled down version of a sunken float paternoster rig that I had designed for pike fishing. Not only could a bait be suspended clear of weed, but if a perch took the worm it could run off without feeling any resistance if the anchoring bomb did not move.

I cast, placing the bait as close as I dared to the lilies, tensioned to the bomb and set the special indicator, of which more later. I then relaxed and awaited developments. Hardly a minute had passed before the indicator flew up and fell to the ground. Line was whizzing through the rod rings as I lifted the rod, which bent, then immediately straightened. I reeled in to discover that the line had been severed close to where the hook had been tied. 'Blasted pike, what with wind and weed that's all I need', I muttered angrily.

Small pike that take baits presented on light perch rigs are prone to give scorching runs. Perch, on the other hand, run off rather more slowly. This is not a hard and fast rule, for fish are never so predictable, but by hitting fast runs immediately small pike can be lip-hooked. The pike might be unwelcome but at least you save on hooks. The next cast produced another fast run. I was ready and landed a beautifully marked gem of a pike – I doubt if it weighed 2lb. Clearwater fish are always strongly marked, and I could hardly wait to behold a perch from the pond.

After two more small pike and another lost hook I wondered if I should try a change of bait – no doubt I could cadge a rudd or two from the angler fishing the other pond. I decided not to bother, as little rudd would only compound the pike trouble. So I persevered with worms, suffered yet more bite-offs and eventually had a fine tussle with a much larger pike that weighed a fraction over 7lb.

Just on dusk the indicator jumped and the line tricked off the reel spool. This is it, I thought, as I waited for some 5 or 6 seconds before lifting the rod. This was better, the fish felt solid, moving quite slowly as the rod tip

'The fish felt solid.'

responded to a series of heavy thumps. The fish was coaxed away from the dangerous far bank and I glanced down to see if the net was handy. Number one I thought, certain that it was a good perch that was tugging so vigorously. Then inexplicably the line slackened. Reeling in only the last few feet, I saw that a tiny piece of worm was adhering to the hook shank – all that remained of a 4-inch lob.

I fished on until nearly dark. The other angler had long since departed so I crossed the path and deposited the larger pike with a number of its lesser brethren in the second pond, remarking as I released them and watched the swirls, that there were no doubt worms in there too!

I could only fish twice for the remainder of the month, but I was plagued with little pike who seemed to find gudgeon even more attractive than worms. All found a new home across the path, except for the only perch, boldly striped with vivid orange-red ventral fins complementing its bronzed olive-green back. That perch so impressive in colouring was hardly impressive in size – perhaps it weighed 12oz.

I did not have to attend school on Thursday or Friday, so in spite of my lack of success I determined to fish the pond from dawn to dusk once a week throughout October. One Thursday afternoon I drove some miles to a small lake, the bottom of which crawled with little gudgeon. They were most obliging and so ridiculously easy to catch that a single maggot accounted for three or four in succession. I must have caught 40 gudgeon in as many minutes. I arrived at the pond just as dawn was breaking. So calm that not a leaf stirred, the water was mirror still except where a shoal of rudd were dimpling the surface. The lightening sky was a uniform grey and it was unseasonably warm. From the experience of my few previous visits I had noted a near perfect perch swim, and had decided not to fish there until I could spend a full day with little chance of disturbance. As it happened only one other angler appeared much later, and to my relief he chose to fish the other pond.

Some two-thirds of the pond was of even width, then two submerged sandbars protruded from either bank, not quite meeting. The gap between the points of the bars seemed to be no wider than about 15 feet, forming a deep channel through which any fish moving from one part of the pond to the other must pass. The bars were clothed in dense beds of potamogeton and even the bed of the channel was overgrown with the inevitable pondweed. There was a dry, sandy stretch of beach within an easy cast of the gap between the bars, with some cover provided by the odd clump of sedge grass. Arriving with the rods already tackled up it was only a few minutes before a paternostered gudgeon was working away, one at each end of the bars. A little cloud of groundbait was catapulted into the channel in the hope of attracting some small rudd, which in their turn might interest any perch in the vicinity.

The lines twitched as the gudgeon tried to seek sanctuary in the weed, but were prevented by the buoyancy of the camouflaged sunken floats. Occasionally a rod tip would tremble as one of the baits made even more strenuous efforts to bury itself. The bright yellow indicators remained immobile in the clips.

Drinking coffee and munching a sandwich I was joined by the six muscovy ducks that seemed to materialise out of thin air whenever I unwrapped any food. They were engaging birds that gave me not only some hilarious moments but seemed to enjoy sitting around my chair. The drake was twice the size of any of his five wives. He was a pompous bird, most of his head was covered in red wattles that gave him the appearance of an inebriated, overweight alderman. He was forever making important-sounding noises, mainly directed at the ducks who totally ignored him. One of his flock was much smaller than the others. She was all white except for a bright red patch around each eye. That duck was quite a charmer and liked to settle down between my feet. After a ritual preening she would engage me in conversation which consisted of muted quacks and hardly audible chuckles. Now and again she would put her head on one side and look at me. Needless to say she conned me out of more sandwiches than were good for her.

The hours passed, the gudgeon had earned their release, and two fresh baits were positioned as before. The duck went to sleep, her head facing her tail and her bill buried deep in snowy feathers. Her crop was bulging, and had she been able to understand I wouldn't have chosen to mention that it was full of duck paté!

The indicators remained motionless. They were designed to show the moment a bait was taken, then fall off the line allowing it to peel from the spool unhindered. Each indicator consisted of a strip of yellow plastic about 4½in long by 1½in wide. A ½in length of Velcro was glued to each end of the strip. The plastic could be bent into a U with the Velcro on the outside. A short piece of bank stick was inserted in the ground directly below the reel, then angled towards the rod butt about 45 degrees from the vertical. A spring-loaded hair grip Araldited to a Terry clip was snapped on the bank stick. Then it was a simple matter to bend the plastic strip, place it over the line and insert the ends in the hair grip. The indicator was then eased back until it would just resist the line tension and the twitches of a lively bait. The Velcro was necessary, as smooth plastic slipped out of the grip too easily.

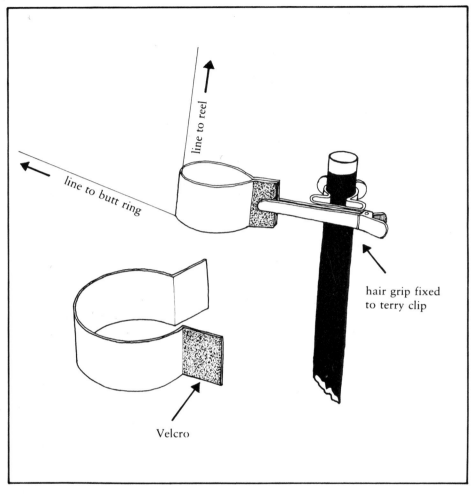

line to reel

line to butt ring

hair grip fixed
to terry clip

Velcro

Vic Bellars bite indication system.

Runs were spectacular – because of the tension the indicator would shoot up, often as much as three feet, and always fell clear of the line.

I glanced at my watch, noted it was noon and realised I had fished for nearly six hours, happily without so much as a little pike to liven the proceedings. I was idly watching the rod tips nodding in response to the still-active baits, when one started more vigorous movements. I hardly had time to realise that the bait was agitated before one of the indicators leapt skyward and fluttered to the ground. The line ran out steadily and as I left my chair hurriedly the ducks scattered, splashing into the water as I lifted the rod, waited for the line to tighten, then set the hook. The fish, ignoring the lilies, headed towards open water. Apart from some heavy, prolonged thumps, and perhaps a couple of surges that pulled the rod downwards I was able to steer the perch towards me. I was certain that I had connected with a large one at last, although I had yet to catch sight of it.

As I eased the fish towards the surface and extended the net, that perch must have decided that enough was enough. It made one surge away from the bank that made me yield a little line, then when that line had been retrieved the fish circled sullenly below the rod tip. Unable to resist the line pressure the perch circled once more, then lay on its side at the surface. I caught a fleeting glimpse of a tiger-barred flank, erect bristling fins and a flash of brilliant red as the perch slid into the net. Perch always look larger than they really are, and this one was no exception, but I was happy enough when the scales registered 3lb on the dot. I slipped that beautiful fish into an

A two-pounder splashes at the net for Roger Miller.

enormous soft mesh keep net. Normally I return fish at once, but in this pond I didn't fancy letting an agitated perch loose so that it could communicate danger to any others.

I cast another bait into the same place, poured coffee and opened the sandwich box. The ducks returned, carefully avoiding the lines where they entered the water, which was considerate of them. The white duck sat under my chair, accepted a further helping of sandwich and normality returned. I wondered, having fished for over six hours for one run, how long I should have to wait for the next. I had begun to think it might be as long again, when an indicator smacked into the rod and fell on the sand. This perch crossed the channel and very nearly gained the safety of the lilies. It's amazing what can be achieved on 3lb line and a soft-actioned rod. Stalemate ensued for a second, then the line, taut as a bow string, pulled the fish's head round and the perch fled for the deeps. The hook held and another perch, seemingly the twin of the first, was netted. The needle on the dial scales steadied on a reading of 3lb 2oz. What a lovely brace, caught within minutes of each other.

I fished on and soon another angler appeared. I asked him if he would kindly take a picture with my camera. His face was a study as I lifted the net and removed both fish. Whether he had ever seen such perch, or for that matter had ever handled a 35mm camera, I know not. The results were abysmal – apart from being slightly out of focus the result of camera shake was all too apparent. It really didn't matter. Later that month I caught another two fish of over 2lb, but it was hard going.

I had a last look round the ponds one November afternoon and stopped to chat to an efficient-looking angler fishing in the shallower second pool. In answer to my query as to how he was faring he launched forth at some length. 'I've lost four hooks,' he said. 'Bitten off. And two fish were grabbed while I was landing them. Can't understand it. It's never happened before – I've even caught a little pike on maggots. There was never any pike trouble in this water, they were all in there,' indicating with a jerk of his thumb the perch pond. 'That's strange,' I said. 'Very strange,' trying to look nonchalant. 'Fish are funny things,' was all I could think of saying as I left rather quickly!

Greenmire Pool

by Roger Miller

That gorgeous midsummer's evening had lent itself totally to the lagoon tench, and it was in Tom's cosy pub after the bat-infested dusk had forced John Bailey and me from the aquarium swim that a regular opened his wallet and placed before us a photograph of poor quality, taken in poor light and slightly out of focus. From three inches to three pounds, there they clearly were, countless perch taken in a single afternoon; countless bristling and magnificent perch all vividly coloured and arranged on the net. The surrounding terrain was quite unfamiliar to us, but a pint of the landlord's best bitter saw the location reluctantly revealed. Another saw access assured and by closing time plans had already been made to visit the pool

A view of Greenmire.

when the surrounding woodland had begun to shade the water's reflections with its autumnal hues.

In the midst of the heavy rains of early October 1987 I stood at the southern corner of the pool for the first time and was surprised to see the water pea-green in colour, like a green mire. So it was with apologies to Richard Walker that I immediately christened the water 'Greenmire', a name that was to prove quite apt as the green-flanked perch it contained were to become quite our friends in the coming weeks.

Wednesday, a dull, dripping, dank sort of a day with rain-laden clouds of slate grey above, witnessed me walking through the wooded path around the lake towards a boy fishing the tiny bay for pike. After days of relentless and persistent rain, whole villages in the county had been flooded. Every pit, pond and stream was filthy brown with floodwater and here was this young optimist livebaiting for pike at Greenmire! With his large bung and snap tackle cast upon the water he bemoaned three dropped runs but, with sail-like, spined dorsals slicing through the surface like the knife through the shower curtain in Hitchcock's *Psycho*, I was elated to realise why this lad was having such a frustrating afternoon.

The young optimist makes it at last.

Another missed 'run' and I could stand it no longer. The instant a shadow of jade and ebony turned on the surface I simply demanded his rod. The pilot float became the main float, a size ten was rummaged from deep within the Barbour and a single swan shot completed the rig. I cast. The float never surfaced. Line slapped against the butt ring as the perch scampered off with the roach livebait, hell-bent on the fallen willow before the hook was even set. The clumsy old glass pike rod was of course quite unforgiving and the spirited fight was conducted almost entirely on the rippled surface. As I grassed this pristine jewel of a fish I was overjoyed that the Greenmire days had at last begun. The boy's reaction was perfect too, he left immediately in disgust at the 'perch problem'.

I scrounged some of his livebaits as he left and I followed him to the car, a mile distant, for my own tackle. Thoughts of the 2lb 8oz perch I had just photographed filled my racing mind. John and I had said we would fish together on the first trip, but before me lay a unique opportunity, an opportunity I had never imagined possible after all the barren perch years. The sun at last broke through as I decided to make a little metaphorical hay. Such was my excitement I had convinced myself I was living in a dream.

Heaven on earth: a country estate, a 2½lb perch and sunshine.

The first Greenmire bag.

It continued as the cut-throats still tore frenetically into the silver, showering sprays of fish. The perch had evidently herded almost every fish in the pool into one corner and I had never known such nerve-jangling anticipation in the sure knowledge that my float would be dragged under at any moment. Round and round it bobbed until it sank slightly and raced sub-surface at an astonishing speed. Again the swashbuckling fight was conducted on the surface and I was soon trembling in unmitigated glee as I slid the second perch of the day into the net. I glanced up and saw that the second float had now disappeared and soon afterwards a 2lb 4oz fish joined the 2lb 10oz perch inside the landing net submerged at my feet. Thus I continued to catch perch that day, until the time came when they simply seemed to 'switch off', leaving behind a scene of utter carnage underwater but wild euphoria above it.

My next visit to Greenmire was arranged with John, and, as is our custom, I arrived at the water first. By the time John arrived I had already missed four takes on carp livebaits in only 30 minutes. This was a completely unexpected development as the perch were evidently seizing the baby carp by the tail and sitting still with them on the bottom, rather like a

The savagery of a feeding perch displayed by this unfortunate livebait.

cat sitting with a freshly-killed mouse and enjoying the anticipation of the imminent feast. That morning I had clearly not been allowing the perch enough time to turn the bait and had consequently continued to strike too early. This simply led to retrieved baits sporting shredded tails and no other apparent damage.

John's first cast produced a very pretty perch of 2lb 2oz that had taken a runt carp well into its mouth and was very securely hooked in the soft, muscle-laden tissue beyond the bone and gristle of the hard mouth. This was indeed a 'red herring' that really threw me off course for a while in discerning what was actually going on down below. We continued to hit takes early in the misguided belief that to let them have the bait a while longer would probably result in a gut-hooked fish, and a gut-hooked perch is a dead perch. I feel embarrassed to have to admit that out of 13 takes that day only two fish were actually landed – the other one by my wife who joined us later that day for a cast or two herself!

The following weekend saw me return alone, as John was starring in midfield for his soccer club in a local derby. I had with me a few tiny roach and dace as well as the obligatory baby carp to use as bait. There was also a new rig to try out that incorporated a flier treble at the tail of the livebait (see rigs section), in an attempt to hook these 'sitting' perch without risking a gut-hooked fish. It worked too! The carp were still being held tail only, with the perch simply 'sitting' with them comatose on the bottom, but the flier did its job. Now the treble was always inside the mouth of the perch regardless of whether it intended to turn the bait in five seconds or five minutes. For a while I was in a position of being able to hit takes early and I appeared to have the upper hand.

As that particular day developed one interesting quirk evolved, in that carp were always taken tail first, as opposed to dace or roach which were more commonly taken head first. My personal interpretation of this was based on these assumptions. The carp, being almost circular in shape, presented the predatory perch with rather a challenge as these powerful little baits had to be caught, incapacitated and eventually consumed. I believed that the perch caught the carp by the usual method of chasing it and nipping at its tail, but due to the vibrancy of the struggling carp the perch could not overcome it at once. Instead, the perch had to sit with it on the bottom and wait for it to exhaust itself in the struggle to escape, and ultimately die. Dace and roach were often taken far more quickly at that same period and it is easy to see why – being smaller, less powerful fish they were easily overcome by the perch and their shape is better suited for immediate consumption. Dace and roach tend not to struggle very much

64

A big perch, the 'biggest' fish of all.

A superbly proportioned Norfolk perch of 3lb 6oz.

The business end of a reservoir perch.

A big perch boils at the net for Roger Miller at a Norfolk estate lake.

The river Wensum perch bend at dusk.

Vivid colours make perch the most striking of fish.

One of Norfolk's biggest perch in recent times: 3lb 12oz.

John Bailey checks the weight of a 'scraper' two.

when lip-hooked and cast and I wonder whether the perch assume them to be nearly dead and take them head first for this reason. Had these baits been available from the very beginning I would have probably caught more perch but learned far less about perch behaviour.

On occasions, just to complicate things, the perch would take roach and dace livebaits head first as usual but sit still with them too. Although a treble was inside the mouth of the perch, due to the bait being lip-hooked, a very tentative hook-hold was all I could ever achieve when the perch were acting in this way. A more effective rig to combat this behaviour was undoubtedly a direct reversal of the flier treble system used with the carp (see rigs section for the actual set-up).

I will never forget landing a 2lb 5oz perch one afternoon, which I had hooked very lightly in this way after the fish had embarked upon a very protracted, finicky take. However, it ended up engulfing the bait with one gulp as it lay on the net. I had left the tiny roach beside its mouth as I took photographs from above. I peered into the mouth of the perch and the roach had been consumed out of sight. When a perch wants a bait it gets it, it seems!

Throughout those early sessions, although I did not appreciate the

A severed dace livebait following a perch attack.

Float, bait and perch rest on the net at the end of the struggle.

significance of it at the time, the water had been permanently very coloured. It was not until the trees began to litter the banks with their falling leaves that the pool began to clear as the colder November days began to clothe the familiar landscape in its winter guise. Soon the bottom became visible over much of the lake and with this the perch simply disappeared. The free-roving livebait rig that had for the most part been so effective was now rendered hopeless.

I felt sure that the perch had migrated away from the dam end with this change in conditions together with the angling pressure I had inflicted upon them. This new-found clarity seemed to change much of what I had learned about the lake and perhaps the key discovery was that the hundreds of baby perch and the more mature roach had also disappeared, probably in an attempt to find some sort of sanctuary elsewhere. With the very light free-roving float rigs I had used up till then a maximum range of 20 yards was all that was really possible. It was then that I decided to change tactics totally and fish at range.

I came down in favour of a well-weighted sunken paternoster pike rig to enable me to present small livebaits at various ranges between 30 and 80 yards, where I knew no bait had ever been cast before. A tackle-buying raid to Norwich was mounted with pike and carp tackle being bought to convert to the Greenmire perch quest.

I arrived the following day at dawn and till three o'clock I sat behind three rods, all fishing livebaits at various ranges of up to 80 yards, all on sunken float paternosters. There had been continuous sunshine since first light but a stiff breeze rippling the water led me to believe that there was a good chance of a fish. Suddenly a large, inky black, angry-looking cloud appeared and obscured the sun for about 15 minutes. The light values dropped very dramatically at that point and the instant they did so the silver foil indicator shot out from the run clip and jammed into the butt ring with the line hissing its way through at an alarming rate. The long-range perch splashed on the surface the moment I hit it and at 80 yards plus it was a very exciting experience. When eventually landed, it weighed a shade over 2lb and, with another small roach cast to the same spot, I soon found myself playing its twin. Mercifully, the perch were taking the baits at once, rendering the timing of the strike easy. The black cloud then passed by, the sun broke through, flooding my world once more with light, and I sat biteless before packing up at dusk with much to think about.

Miller sits it out!

The clear water conditions persisted and I continued to pick up odd fish by casting around the lake. The light values proved their importance once again on a day of direct contrast to the one I have just described. A very dark, wet and windy November day saw me huddled beneath the brolly in mild wonder that a run had not been forthcoming when suddenly a gap in the clouds appeared from nowhere and the sun burst through. The light values *increased* appreciably on this occasion and in the few minutes of sunshine that day I was rewarded with a fine perch of 2lb 3oz. On both occasions the *change* in light values had induced the perch to feed. Overall, however, the sunny days with wind in the northern quarter, coupled with a spell of high pressure, would see the Greenmire perch very hard to tempt. With other projects on the go where this kind of weather had less of an adverse effect on the fishing, I began to study the weather forecasts much more seriously before arranging trips to Greenmire.

The mildness of much of December 1987 brought with it the realisation that I had to fish Greenmire hard whilst the weather was being so conducive to the campaign. The welcome torrents of rain returned the greenness to the water, and the perch, as I had hoped, returned to their former beats just off the dam wall. I continued to paternoster baits, but brought them in close where I had enjoyed the earlier successes. The heavy rains and roaring winds brought with them a new problem to contend with, namely under-tow. On occasions the undertow would actually pull line from the strongest of clips. It also seemed likely that the sunken float of the paternoster rig was being pulled down in the water, therefore allowing the livebait to conceal itself on the bottom amongst the mass of fine silkweed. On one particular day the undertow actually dragged each set-up, including bombs of up to an ounce, out of position. The underwater flow was always moving in the opposite direction to the wind, which persistently blew from the south or south-west. The lake was virtually unfishable at this point so I decided to revert to free-roving baits again in the hope that the two opposing forces on the tackle, that is, wind on the float and undertow on the bait, would negate each other. I was desperately keen to present a natural-looking bait in these conditions and the idea did seem somewhat outlandish, but, amazingly, it worked! The float remained rock solid with the roaring southwester above it, a powerful undertow beneath it and the bait was not dragged one way or the other out of position.

It was at the precise moment that I began to fish in this manner that perch began to return to my waiting net. The perch were evidently milling around the dam wall again and I was expecting great things. As before, on some days the baits were being engulfed immediately and on others the maddening tail-nipping habit returned. By this stage the flier treble proved to be the kiss-of-death on a bait and any perch taking a bait and feeling the metal of the treble would eject it immediately. The solution was obviously to get rid of the flier, but the old problem of when to strike remained. I was convinced at this stage, however, that there was no danger in leaving a perch holding on to a bait, no matter how long, whilst it remained dormant on the bottom with it.

My new approach was to ignore the initial take and wait to see whether the perch scampered off with the bait or whether it just sat with it. In the latter case I resolved to stand by the rod and force myself to strike only when the perch embarked on an actual run. I have in fact had to wait by such a rod for up to ten minutes before eventually striking as the perch at last moved off with the bait. I have even had an initial take on one rod, and been waiting by it only to have a take on another. I have subsequently gone through the usual procedure with the second rod, landed the fish and still had to wait for a strikable run on the first rod.

At this time a new supply of livebaits had appeared in the form of about 100 stunted little roach. A grossly overstocked little water had been thinned out a bit and I was pleased now to have a ready supply of roach livebaits at the bottom of the garden. The now solely roach livebait stock proved to be a

A Greenmire Pool perch taken on a 150-yard drift at dusk.

70

better investment than the carp, as far less trouble was experienced in using them as baits. Runs were more positive and bites easier to hit but I always felt that the bigger perch preferred the larger carp.

These roach continued to account for a number of 1½lb to 2½lb fish that came steadily to my rods in the period leading up to Christmas and it was at this stage that I began to wonder about the possibility of a lone perch or two, far larger than I had already caught, staying out in the middle of the lake, aloof from all the rest and uncatchable by my current methods.

One evening during this period I attended the monthly Norfolk Anglers Conservation Association meeting where local pike expert Dave Batten was giving a talk on his refinements of the original ET drifting rig invented by Eddie Turner for pike fishing. Following Dave's excellent talk, he and I discussed perch fishing at Ardleigh Reservoir in Essex, a subject I knew nothing of whatsoever. He told me of one of the pike men who flooded into the place after the record pike was caught and of one in particular who had caught a 3lb perch on a conventional ET drifting rig, intended for pike of course. Now I did not believe for one second that the Greenmire perch would tolerate anything quite so coarse as a standard pike drifting rig but as soon as I returned home that evening I got straight to work in the shed and knocked up a miniature drifting float illustrated in the rigs section of this book. It only takes a few minutes and can be easily made using items already in the tackle bag and a pair of scissors. I really do not know why I had not thought of it before. I am mildly surprised that fast Eddie has not brought out a smaller version of his float to cater for a more refined type of drifting for pike as well as the ever-growing perch market.

The following day I went drift fishing at Greenmire with my much modified little drift float and, as in pike fishing, whole new horizons could be opened up with a real chance of tracking down the king perch in Greenmire's castle. Fortunately the south-westerly that had been blowing all week was still rippling the lake, a drift fishers dream! As can be seen in the diagram I operated from two extreme vantage points. By alternating between the two at regular intervals during the drift the float would kite across the wind as shown, thus criss-crossing the entire lake.

Up until that day my whole set-up had been geared to perch and nothing else. Pike were so thin on the ground that I had only ever caught two in about twenty trips, and I had moved both of them to a nearby pit where they would grow big and fast. I had therefore persisted throughout the campaign with the 4lb line tied direct to a size 14 treble. The little drift float worked every bit as well as the bigger versions and goodness knows at what range the bright red square at last disappeared. Eventually I made contact with the

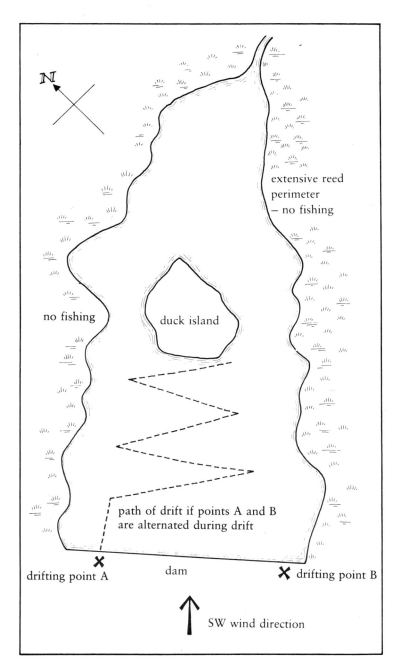

Greenmire Pool.

fish but never gained an inch of line. After an utter farce that lasted at least fifteen minutes the enormous pike crashed into the reedy perimeter of the lake sending the coots wild with panic. I eventually recovered the float and the rest of the tackle intact, thank goodness, but from that point on I was unable to justify not using wire, and, as I expected, the inclusion of the wire in the rig simply meant hardly any perch. To revert to 4lb line for drifting would put that big old pike in too much danger for my liking so I very reluctantly forsook the continuation of drift fishing at Greenmire, even though the method was every bit as effective as it is for pike.

As the excellent weather continued, so did the steady stream of specimen perch, but as this is a species with quite rapid learning capacities, I was never able to settle into a routine and just enjoy each fish as it came along, rather I was continually striving to discover their latest quirks. One quite astonishing breakthrough came during the last fifteen minutes of a particularly dour day. The baits were tired and lethargic and I decided in the twilight to cast three fresh baits to the same areas in the vain hope of inducing a take. As the first bait hit the surface it was taken in an instant with a scintillating run that resulted in a lovely, pristine perch of 2lb 4oz being landed. I then managed to get two more baits out, both of which were taken within seconds of being cast, and fish of 2lb 2oz and 1lb 15oz were landed. It was clear that the perch were homing in on the baits as they splashed on to the water and if they were not taken within a few seconds I knew that they were going to be ignored. Indeed, I left one such bait out throughout this spell only to reel it in at the end of the session totally unscathed. The other two were being taken at every new cast. I finished up with seven runs and five perch of weights up to 2lb 5oz.

This new trait was something I exploited with reasonable success in later trips, but the mortality of the livebaits was very high and the feeling of unnecessary slaughter pervaded every trip in which I kept casting baits throughout the day. I hoped that dead fish would induce the perch in the same way but unfortunately they did not. Indeed, not a single perch ever came out on a freshly-killed deadbait from Greenmire, unlike other waters discussed in this book where it was an absolute killer.

I was pleased that these induced takes, if that is what they actually were, did not see the perch sitting with the baits again, not for a good while anyway, and when the tail-nipping resumed I tried a piece of advice offered by John Wilson. This was to gently lift the rod and feel down for the perch and draw very lightly on the rod tip. Sometimes a violent take would result and on others the bait would be rejected. My former method of simply waiting for the eventual run appeared more successful on this water.

More success at Greenmire.

Mrs Miller too could catch at Greenmire.

It is now mid-January and the weather is still mild, wet and windy, but, as I write, new and exciting projects are waiting to be exploited and as I reach the end of this section on Greenmire I realise that I shall not return this season. Now that I have written its story I feel that my relationship with the place has reached its natural conclusion.

Greenmire will forever have a very special place in my angling memories for several reasons: the continual development of my approach to fish that were never easy but which responded to logical advancements of tackle and technique; the frenzied excitement of running between rods striking like a madman when the perch decide that they really want to feed hard. But perhaps my most cherished memories of the Greenmire days are the recollections of those magnificent dorsals slicing through the surface film in their buccaneering defiance of my jagging rod, when only the daft old coots, bobbing like buoys in the early morning mists, were my witnesses.

The Reservoir

by Nigel Witham

A young angler who burst on to the angling scene with mind-blowing specimen roach to 3lb 14oz from his local River Beult in Kent, Nigel has of late, however, spent a considerable amount of time in the pursuit of perch from large reservoirs with superb fish to 3lb 9oz being landed in the process. Nigel's contribution to the book centres around his lure fishing experiences. He is very much a thinking angler who has no preconceived ideas about his quarry but who holds dear all the true values of angling without losing his innovative approach.

There is a water not far from my home where fishing legends have been founded. Not just once or twice, not a few times, but many times. Imagine a water of hundreds of acres, a confluence of three flooded valleys shaped so that if you unfold an Ordnance Survey map of the North Weald it appears like a crow's footprint stamped on the land, straddling the area from the edges of the little village of Wadhurst to the vineyards stitched on the sides of the outfall valley near Lamberhurst. Sunk without trace beneath its near hundred-foot depths lie petrified hedgerows, knots of trees, the gulleys of old streambeds and ponds, and dotted here and there I am sure are the part-preserved ruins of ancient farm buildings. I describe Bewl Water and it is very, very beautiful. It is also a premier trout fishery – sometimes I have thought it is run by men who care only for fish with an adipose fin and know all others as vermin and pretenders to their water.

Up until 1982 my friends and I had caught many fine perch but then Bewl was opened to coarse anglers for eight days during October and November. After that we had more than an inkling that our perch prospects were improved considerably. They were, and later, had we been keen to publicise and run the gauntlet with the water authorities, then I am sure our catches would have made history and we would have all enjoyed short-lived heroic notoriety in the angling press. Let me tell you more.

The purpose of allowing coarse fishing at Bewl was to help the removal of pike which were to be used for the stocking of other waters. Netting had only been partially successful. The problem was that pike and trout work

Steve Burke of 'The Perchfishers' with a magnificent perch of 2lb 15oz from Bewl Bridge Reservoir.

well together and the pike were big – some were very big. Unfortunately the pike and trout anglers did not mix and it only takes a few bite-offs to cause a bailiff's hackles to rise like a brownie taking crane flies.

Whether the cull was appropriate is now very dubious but at the time it drew keen pikers from all over the country. Late October mornings at Bewl had never experienced the like, as a hundred men sped away to the favourite spots, outboards spluttering, leaving behind a peculiar blue haze. Two-stroke fumes draped the mist, gaseous oil slicks hung about the jetty tracing the course of the eager anglers before dissolving away into the damp air. During that first year many pike were removed: thirty-pounders; many fine doubles for a day's effort; twenty-pounders were reported on most days, and many I am sure were surreptitiously slipped back. Amongst the pike were other intruders, stripy pirates of the trout seas. Perch were present in numbers and whilst not quite two-pounders another season's growth would see to that. Luckily not many noticed their presence.

The following season we were back: Steve and Stewart, Martin and I, quietly perching whilst hoping to keep our secret from the many pikemen

A fine brace for Nigel Witham; 3lb 9oz and 2lb 11oz that fell to legered lobworms.

and unimpressed bailiffs. Others too had come for the perch but like us they did not want their true reasons for being present revealed, and so it was some time before we met and were able to gain from each other's experiences. Later, it was from this group that the Perchfishers were formed.

I shall tell a little of our catches later but for now I will attempt to explain some fundamentals of perch behaviour that any reservoir or big lake anglers should be aware of, if they are to succeed. There are theories in angling which are improperly labelled such because they are not proved. Of the following, however, I am convinced and our catches bear testament to its correctness. Since Bewl our methods have rarely proved unsuccessful on other waters.

Most anglers know that in any deep water during winter, temperature stratification occurs and a point is reached where the deepest water is the warmest. In our climate this does not always happen and indeed, as I write, during the early part of February 1988, the local conditions have not been severe enough and generally water temperatures have not dropped below the critical point of approximately 4°C (39.2°F). From long-standing observations it is apparent that most of our native species of fish prefer to live in the warmest water when overall conditions are at their coldest. This is true of perch. I don't think we need examine the reasons in further detail here, but suffice it to say that in the middle of winter you will find shoals of fish in the deepest parts of your waters. Please note that I am concerned here with deeper waters and that shallow ones have a slightly different set of rules.

The perch is peculiar amongst our common species. It is one of the most highly developed and has evolved to have a swim bladder that is remote from its oesophagus and can only be inflated or deflated very gradually by means of gas transfer, which has to take place via its bloodstream. This means that the perch can comfortably change depth by only small amounts at a time and, furthermore, the shallower it is swimming the more difficult adjustments become.

In summer, perch will be found hunting amongst shoals of fry in the shallow margins of the water. In winter they will be as deep as they can get – if possible, many tens of feet. As the fish can only alter their depth by significant amounts gradually there is a consequent slow migration of shoals from shallow to deep during the autumn and a reverse migration back to the shallows during the spring. This latter migration should not concern us as it occurs chiefly during our closed season.

Nigel Witham's personal best perch of 3lb 9oz.

Migration takes many weeks and I am not completely certain when it begins. It is well under way by the beginning of October when fish will usually be found in 10–20 feet of water. We found when fishing at Bewl Water that individual shoals vary slightly in behaviour. There is no apparent reason why some shoals were found to be more advanced than others in their movements. I am not sure whether a perch can tolerate sudden changes in depth away from its preferred level – whether, for instance, a fish swimming at 20ft can change depth by plus or minus 4ft (20 per cent) without feeling undue discomfort, or whether it can adjust by more or less than this amount. Doubtless it will be most comfortable at the depth where it has neutral buoyancy and we must expect to fish for it there. Hypothetically, this depth will vary according to individual fish, but I suspect that the total variation within any shoal is quite small – I would hazard a *guess* at 10 per cent of the water's depth.

Other locational factors must be considered, though I don't think them as important as the depth/time-of-year equation. Camouflage is vital to the perch, as can be deduced from its distinctive coloration, and accordingly you will seldom find *feeding* fish far from the bottom, whether this be the bottom at the 10ft or the 100ft contour. Better still are sunken trees, gravel bars, and, more specifically on trout reservoirs, fish cages – which I assume are favoured because they cast a shadow below and because the frequent supply of food attracts small bait fish.

To recap briefly, in summer the fish will be within the first few feet of the water's surface, in winter on the bottom and in autumn and spring gradually changing through points between. The perch has to be an opportunist, living, it would seem, in a state of constant flux.

Another major factor which affects greatly the activity of perch is the level of underwater light. I am very careful to use the phrase *underwater* light because many influences affect the level of subsurface illumination which some anglers may not appreciate. A bright day above the surface does not always imply one where the fish swim.

I used to take a light meter fishing – not a normal photographer's type, but one connected to a long probe which could be set six feet below the surface. I took measurements and recorded all captures for a whole season whilst fishing for both roach and perch. Because there are many variable factors you need to catch many fish over a long period to have results which bear statistical analysis. An angler fishing for one season cannot hope to draw accurate conclusions. However, there was an apparent correlation between light intensity and frequency of capture which varied only with the size of fish. Small perch feed most of the time; big ones when they choose.

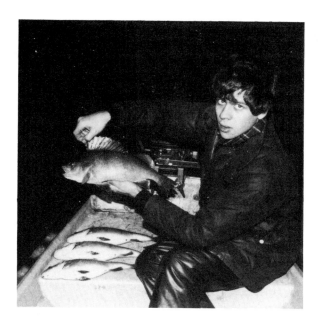

Nigel Witham with a splendid bag of Bewl perch of 2lb 13oz, 2lb, 2lb 7½oz, 2lb 10½oz taken whilst spinning.

When big perch, say fish over 1lb, choose to feed it is when they have most advantage over their prey. This will not be in bright conditions because the advantage cannot be great enough when the prey can see the predator. Nor can it be in darkness because no fish can see when there is no light. Perch feed best in intermediate conditions and zoologists confirm that the sight of perch is biased to enable them to do so.

In very deep water the light may only reach suitable levels at midday and in clear, ripple-free conditions. We would expect high pressure conditions to provide us with the best sport during the middle of winter. In summer it is well known that dawn and dusk are the best times for perch fishing. In autumn, the time of year when my friends and I have done most of our perching, we have found that the duller the weather, or the more chop, or the cloudier the water then the closer to midday we would expect the fish to feed. Conversely on bright/wind-free/clear-water days, the closer to dusk the better. On Bewl a four o'clock curfew was imposed and we all caught more after the clocks were set to GMT which gave us an extra hour at dusk.

We have decided that at certain times of year the fish are at given depths and that some shoals are at slightly varying depths to others. We have also decided that perch feed best in specific light levels. There is a clear link. Shoals of fish swimming at different depths will feed at different times of day dependent upon the light, which varies with depth. The amount of ripple or water colour also matters but these factors normally affect all shoals equally.

Of course all this is only relevant in autumn when the shoals are spread at various depths. However, if you consider that autumnal conditions, as far as the perch are concerned, constitute about half of our fishing season then you will see the relevance of my remarks. From September to December you can expect to catch fish from intermediate depths. What is more I would add that I generally do little perch fishing during the summer months because it is so unproductive – due, I expect, to an over-abundance of natural food caused by large fry populations. Whatever the cause, I no longer consider summer perching a worthwhile occupation.

From the above we can predict that Person A fishing in, say, 20 feet of water during the early morning might expect to stop catching at midday, whereupon Person B fishing in 30 feet begins to catch until the light drops. Person A then gets good sport until dark. Perhaps you think that I have drawn too many conclusions, but I can tell you that the above scenario happens exactly in reality. It has done so for us many times.

I regard methods of fishing as of very secondary importance. At Bewl we were limited to artificials or sea fish deadbaits over 5in in length, which left us few options. From our early experience we can rule out the deadbaits without any discussion so let's start with spinning.

I enjoy spinning very much and do a good deal of it. Most observers of perch fishing would claim that spinning was a comparatively unproductive method, but I disagree. I look at it this way: I have an enormous water within which the extent of the perch population is uncertain. My starting point is that I have an inkling of which depth to try. I don't expect to see the fish except possibly by the use of a sonar device – but even this is dubious when seeking perch, because whilst you can easily spot a large pike, a shoal of medium-sized fish may be of one of many species. The best means of locating fish is to hook one, and the more water covered the more likely contact becomes.

Trolling is the most efficient method, followed by casting a spinner either from a boat or the bank. Mobility is clearly essential until you find favoured spots and the lure angler can be very mobile. If I were fishing a new water I would probably start by trolling. This is a method that enables you to fish at varying contours by simple adjustments in tackle, and so by using two or even more rods you can experiment confidently whilst fishing. The chief disadvantage I have found is that it is difficult to troll in depths over 20ft or so. For perch a very slow speed is often required. Anchoring from time to time, in likely-looking spots, allows such areas to be fished more carefully.

A misty dawn on Bewl saw Nigel Witham tempt this marvellous perch to a deep trolled lure.

Very deep water can also be covered best whilst stationary. Clearly a boat is a necessity when trolling, and an electric outboard motor and echo-sounder very desirable assets. Bank fishing does work because I have seen others succeed, but I do not contemplate reservoir perching without the use of a boat. Fortunately this has always been possible. Incidentally, all the boats at Bewl were fitted with petrol outboards but these did not seem to worry the fish.

Lure selection is a complex matter to some and a simplicity to others. I am firmly of the latter category, although in fairness to friends I must add that there are others much more qualified to comment than I. The problem with lures is that once you gain confidence in a particular type and colour it becomes a matter of the greatest reluctance to effect a change. Accordingly, objective comparisons between different types of lures are difficult. My favourite is the Abu Atom spoon in the largest size, and in green or blue scale finish. I have about a dozen of these, all identical, along with at least 50 others in various sizes and colours, all of which I have acquired because every season I am told by one tackle dealer or another that they are about to be discontinued.

Another favourite is the banana-shaped deep-diving plug generally known as the Flatfish. Trolled at the right pace and distance these will go down to about 17ft without lead-cored line. A 5in chromed model has been an outstanding success for one of my friends, who has on four occasions caught two perch simultaneously on the one plug. Imagine that next time you look over your local trout fishery: a brace of perch of nearly 6lb hanging from one plug. Unfortunately the chrome version is currently only available from the USA. Some members of the Perchfishers jointly own one rather unshining, well-dented example and are currently considering offers!

Of course many other lures also work. I have known fish come to Buzzers (we had them in 1983), Droppens and other bar spoons, Big S plugs and Tobies. I have no doubt that many other lures would work if put amongst a shoal and fished at the right speed. Bacon rind and other additions such as red wool dipped in concentrated flavours – try Rod Hutchinson's salmon – sometimes help. My best lure-caught fish of just over 3lb came to an 18g Toby.

It may sometimes happen that when two anglers have been fishing exactly the same water, indeed from the same boat, with exactly the same lure, their catches may be markedly different. This can be most frustrating if you are the suffering partner. I can recall two consecutive days of fishing with Stewart Allum when I was thoroughly thrashed and there was no apparent difference in our technique. We even swapped ends of the boat. I can offer no reason, but if this happens to you then I can assure you that your experience is not unique. Ten 2lb perch versus one can be rather an embarrassing result, especially if it is not a one-off occurrence. Mind you, I've had it my way too.

Wire traces are optional but recommended. They won't worry perch but will mean that you should stand a better chance of landing any pike. I always use one. Other tackle is very much a matter of personal choice. I use 6lb breaking strain line because there is no need to use lighter. When there is no need to cast far I use stronger. A light spinning rod is most fun and I have an 8½ft through-actioned glass model with a winch fitting and five enormous single-leg fuji rings which I enjoy using immensely. It has a test curve of about 1¼lb.

Once you have located your fish you can then contemplate other methods of capture, and all standard techniques apply. As in all fishing simplicity is often the key to success. I recently shared a very enjoyable day's fishing on Ardleigh Reservoir with Stewart Allum. We knew the spot, adjacent to a trout cage, from the previous weekend. We simply anchored, dropped a couple of pints of maggots over the side and fished worms lowered beneath the hull with no more than a hook and a swanshot on the line. Bite indication was by touch and the rods were rested on the gunwale. By dusk we must have had over 50lb of fish, mainly in the ½lb class but with a good number over the pound and up to a shade under 3lb. This is quality perch fishing. I lost an enormous fish which I hooked on a livebait, a mistake I am not proud of. As an example it serves to illustrate the secondary importance of bait selection. Livebaits, deadbaits, worms and maggots all work and generally the bigger the bait the bigger the fish. I have seen a perch take a rudd of nearly a pound. It was the biggest perch I have ever seen and it was never caught.

Bringing perch up from great depths is a problem. It causes the fish much distress because, as the pressure decreases, the gas in the swim bladder expands and increases its buoyancy to a point where if the fish is tired by a long fight it is unlikely to be able to swim down again. This means that the fish will die.

The solution is to get the fish up quickly and release them immediately. Photography, if any, must be minimal. I have few pictures. Many almost certain three-pounders were not even weighed by us. At times we released fish and watched them fin vigorously straight down only to appear helplessly a minute or so later at the other side of the boat. All you can do is sack them and sink them to the depth from which they were caught. This allows them to recover prior to release. Unfortunately a few casualties may be inevitable. The Americans have ingenious devices which claim to set the fish free at the appropriate depth. I like the look of none of them, but all improvements in fish handling are to be welcomed. Again, fishing from a boat gives the fish a better chance. I don't think many of our fish suffered, at least not once we knew of the problem. Please take care, as these problems are certain to occur when fishing over about fifteen feet deep.

I suppose that during six days of fishing on Bewl Water in 1984, and between five boats, we had approaching 200 perch of 2lb or more. Three-pounders represented about one in ten caught, with a best of 3lb 12oz. I remember taking three 2lb fish in three casts in 1983, and catches of ten fish for nearly 30lb came to someone on most days.

After such excellent sport we had high hopes for bigger fish in 1985. Tragically our optimism was misplaced. Three days on the water and we had not even pricked a perch. The pike had been reduced to a seething army of 2lb jacks. I had an eight-pounder which was one of the best caught. The sole high point came when I connected with a near 6lb rainbow trout on my light tackle and took 30 minutes to boat it after a series of turbo-charged runs and Olympian high jumps.

Bewl perch being played, . . .

. . . drawn to the net, . . .

. . . landed, . . .

...weighed...

...and returned.

Day four was pleasant enough, civilised weather for early November. The sun obliged, the wind sulked quietly and the enveloping forests turned from soggy brown to deepest russet, and finally to planished gold, transformed by an alchemist who had the use of no device other than the magic light from the climbing sun.

Lunchtime came and, feeling civilised ourselves, we pulled out and beached the boat on the edge of a shallow, gravelly bay so that we could eat lunch comfortably. It had not rained for some days and the soil beneath the scrub line was powder dry and dusty so that as we trudged up the gentle incline our moonboots left hardly any impression at all and sent fragments of crumbling clay plopping down into the water. Only this and our heavy breaths caused by too many layers of warm clothing broke the hushed stillness. The earth was too silent, like an evil child about to be found out.

4lb 3½oz – a magnificent perch in every way.

Stewart saw it first: the rotted carcass of a good perch, its body cavity half-filled with sintered soil, its upturned ribcage jutting up and out. Weather-bleached bones were exposed like the long-expired skeletons of parched desert mammals. It was just the shell of a fish, and there were others – we wandered a hundred yards and saw a thousand more. We ate lunch in silence, heads hung low. The trees were no longer dressed in garments of precious metal but fool's gold.

We returned and fished a little longer, and as if to tempt us the water threw us an anomaly – a perch of 1¼lb. We packed up and left. I have known temptresses other than Bewl and to this day I have not returned. My back is turned on the place. To fish for ghosts is to destroy your own soul. For Bewl perch the future does not look good but the disease has run its course now and tons of fish have perished. Perhaps our lone survivor will be the precursor to a recovery and will help conceive new legends. If so I will return. Meanwhile there are other venues to fish, many I am sure with unknown but great potential. I shall take advantage elsewhere. The perch is a magnificent fish. I shall see many more yet and they will crown many fine days for me. I am determined that this shall be so.

The Use of Fish Recorders on Large Reservoirs

by John Nunn

John is a good friend of both authors, and they have spent many happy days fishing together on a multitude of different waters. He has a very impressive list of specimen perch reaching as far back as the late 1950s, and has been one of the few anglers to have exploited the use of modern fish-finders on large waters. His fascinating discoveries form the basis of his contribution to the book together with the tale of his personal best perch of 3lb 4oz taken whilst using his Humminbird fish-finder. John is a fine all-round angler who lives in an idyllic home on the banks of Ormesby Broad.

'It is in those dark murks of water that are so fascinating in themselves that you find the perch'. This is Mr Crabtree's viewpoint on perch location, and he was never far from the truth. His advice is still worth following when trying to locate perch. They do love the 'dark murks' and seem to home in on features to be found in these areas. My recent experiences on a large, deep reservoir have strengthened my attraction to the deeps.

To be faced with over 100 acres of deep, unknown water is a daunting task. And travelling for two hours along frosty winter roads, will make any fisherman even more determined to get it right. The angler of the 1980s has sophisticated fish-locating equipment to turn to for help, in addition to the watercraft taught us by the likes of Mr Crabtree. One such locating device is the Humminbird – an American import. A screen displays details of depth, but more importantly it also shows the presence of fish at certain depths. You can even magnify a specific depth band by adjusting the device to zoom in on a chosen depth, thus giving greater detail. Shoals of fish, even individual large fish such as pike and trout, are clearly displayed. The more you use the device, the more you learn from it. Features of the lake bed such as sunken trees, abrupt depth changes and even variations in the hardness of the lake bed show up clearly.

With such sophisticated equipment at my disposal for the day, I rowed

An arm of the reservoir.

out on to the reservoir, excited by its considerable reputation for large perch. As you might expect in winter, vast areas of water proved to be devoid of fish: frequently areas that looked right and had attractive depth variations were places where it would have been easy to waste long periods of the short winter day.

First indications of fish were seen in 30 feet of water, where a number of snags showed up on the reservoir bed, presumably tree stumps. As I progressed along the reservoir arm, the correlation between submerged trees and shoals of fish became more distinct. I fished alongside the stumps, taking perch from each area that was shown to have a congregation of fish.

The reservoir is used by a sailing club for racing. They mark their racing courses with permanently moored buoys. These buoys and their ropes form the only features in an otherwise featureless area. Hundreds of yards of empty water, then a single rope, and around that rope: a congregation of fish! After fishing around these mooring ropes and catching large numbers of perch of up to 1lb, I decided that an experiment was needed.

I rowed over to an area of the reservoir which the locator screen showed as having no fish in it – a difficult thing to force myself into doing, but it was in the interests of scientific research after all. The nearest concentration of fish was around one of the trout cages some 50 yards away. I lowered the boat's mooring weights to the bottom, some 32 feet below me. After 50 minutes the Humminbird showed the arrival of a few fish close to the bottom. Sure enough, bites started to come and fish came regularly to the net. Sport became hectic, and a glance at the screen showed why – a large shoal had collected underneath the boat, seemingly attracted by the mooring ropes and weights. I had created a feature for them to home in on.

Excited as I was by the success of my experiment, the size of the perch was disappointing, with no fish topping 1lb. Up to this point I had been fishing paternostered livebaits on one rod and large lobworms on the other. A few days before my visit a fellow Norfolk angler, Dave Batten, had achieved success with a small lure jigged in the style of mackerel feathers, beneath his boat.

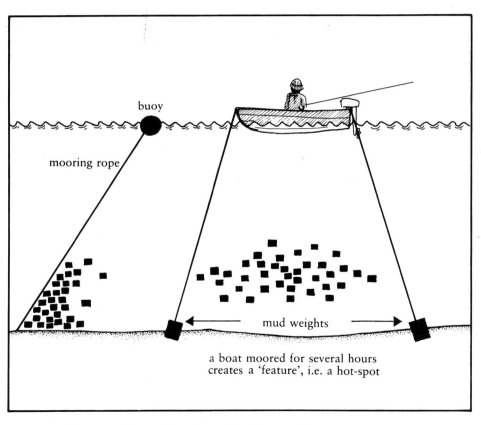

buoy

mooring rope

mud weights

a boat moored for several hours
creates a 'feature', i.e. a hot-spot

Reservoir hot-spots for perch as shown by fish recorders.

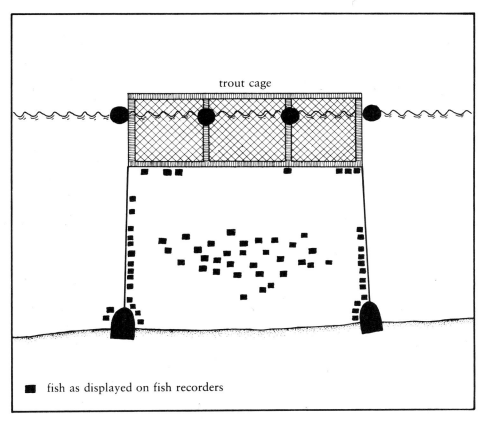

trout cage

■ fish as displayed on fish recorders

Perch hot-spot under trout cage as shown by fish recorders.

I searched through my lure box, finding an ancient quill-bodied 2in Devon Minnow, a lure given to me by my elderly neighbour – a gentleman who had once owned a tackle shop in Derby. In order to fish such a light-bodied lure in that depth of water I added a small barrel weight, which I stopped with a swivel six inches from the lure. The Devon's armoury is composed of two tiny trebles. Allowing the rig to sink to the bottom and watching the line go slack, I took in three turns of line before starting to jig the lure. The response was immediate, a violent take followed by a lively struggle. Soon an immaculate perch lay in the net – at 3lb 4oz my day was made.

Further fish followed, a 2lb 14oz fish adding to what was already a memorable day. Perch remained under the boat until I packed at dusk – the digital glow of the locator reminding me of its role in the day's sport.

John Nunn with a cracking perch of 3lb 4oz from Ardleigh Reservoir.

I rowed back reflecting upon the reasons behind the pattern that the fish locator uncovered. I am tempted to say never mind the reasons – at least if we know what *does* happen, we can make the most of that knowledge. Nevertheless, the thinking angler should be asking questions. Perhaps the perch find something to eat on the ropes? But as the boats' mooring ropes spend over half of their time out of the water, what possible nutrition can be found on their man-made fibres? I would expect the mooring ropes of the trout pens and the sailing club buoys to be possibly more attractive. The perch may graze them in the same way that sea fish may be observed grazing mooring ropes in a harbour.

I am not keen on the suggestion that the boat acts as an attraction. What difference could it make to life 30 feet below? I purposely did not put any maggots or worms around the boat to attract fish, so that the experiment could not be influenced by such an obvious man-made attraction. I have spoken to others who have fished the reservoir and they have all commented, without any prompting from me, that as their day progressed they caught more fish from the area immediately next to the boat. So, the ropes would seem to be the attracting feature. At least you can set out confident that casting is unlikely to be important – just lower your bait over the side!

The Police Diver's Pit

by Roger Miller

The Norfolk Police Diving Squad is, naturally, the best in the country and one of them, a colleague for many years, occasionally sidles up to me and tells of fish sightings that he tantalisingly phrases as 'something that may be of interest to you'.

Two years ago when he made such a remark it transpired that, diving the day before, he had seen some perch in a two-acre gravel pit of quite a depth. I quizzed him further and his rough estimation of their length suggested that they were of a size that was very interesting indeed.

He personally found them fascinating, for virtually all the other fish he encounters underwater are at worst simply curious of him. Tench, bream

The beginning of a perch quest? (Photo by kind permission of the Eastern Daily Express.)

and pike commonly ignore him. Roach and rudd are wary but untroubled by him as he gently eases across the bottom amongst them in some gloomy, watery hole in search of safes, bodies or murder weapons. However, big perch are terrified of him. They are unquestionably the shyest fish he has ever encountered. As soon as big perch see him, from whatever distance, they flee in panic. The perch seen the day before were no different.

This intriguing observation was compounded by another diving friend, well-known angler and tackle dealer John Wilson, who has also witnessed big perch being very, very aloof and incredibly easy to spook. Big barbel could be cuddled like babies underwater, it seemed, and enjoy it too, but there was never any chance of getting anywhere near a big perch.

The first session at the pit was full of promise – a mild, overcast day saw me forging a path of sorts through an indescribable tangle of bramble towards a shower of silvery fish betraying the presence of a mammoth perch. A free-roving livebait cast to the spot together with a float-paternostered bait by the deeps began the campaign, but for two hours I sat runless.

Suddenly the silver paper fell from the clip. My hopes were low, for the

Where's the hot-spot then? (Photo by kind permission of the Eastern Daily Express.)

The Diver's Pit in all its glory.

high wind and the slack line convinced me it was not a fish. I left the bait where it was and snoozed the afternoon away.

Dusk began to fall and with spirits low I reeled in the baits. The paternostered bait that I believed had not been taken, had a tiny scratch on both sides of its tail root together with a number of missing scales from the same area. Everything indicated a perch attack. If only . . .

Again, as on the first visit, a big perch was attacking small roach in the back bay and with a tiny float supporting a free-roving roach I commenced the session with a new optimism. I never raised a perch all day. My confidence was evaporating but two young anglers fishing an adjoining pit about the size of half a tennis court showed me a perch of about a pound in the net. They estimated it at about 12oz, a very encouraging estimation in view of them telling me of a 2½-pounder and even one of 4lb caught there earlier that season.

This knowledge gave me a big boost and I stuck with that pit for quite a while — lives, deads and lobs covered every inch of that little water but I suffered blank upon blank. Occasionally, as if to taunt me, a big one would splash on the surface and then disappear for the rest of the day. No matter what I tried the perch never responded.

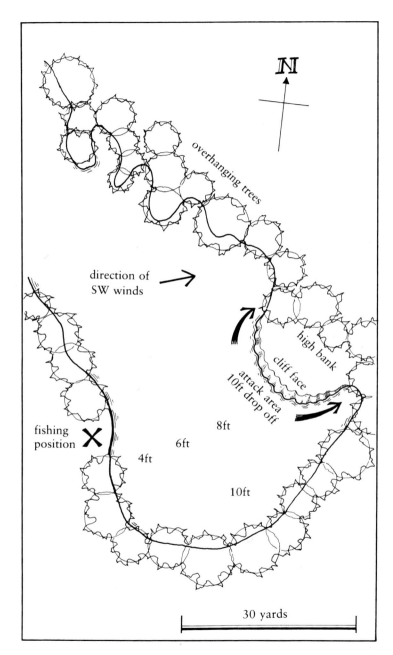

The Police Diver's Pit.

I again conversed with the local expert juveniles. Yes, they said, the perch previously caught were genuine. I had only fished on Sundays and apparently the pit was used as a 'practice' venue by the local match club on Saturdays, due to the multitude of small roach that inhabited the place. The matchmen would introduce vast quantities of maggots in an attempt to keep as many roach in the swim as possible. Very occasionally a whacking great perch would come out on a 1.1 bs bottom tied to a size 22 with a single maggot as bait.

I was told I would never catch them any other way. Some of the matchmen had tried worms but the perch wouldn't look at anything other than maggots on fine line. I considered this as a new approach, but God only knows how many baby roach would have to be caught before one of the big perch picked up a bait. I tried building up a roach swim and fishing with live and deadbaits but again to no avail. Worms were pecked to pieces by the roach so I gave up on the place.

Maggots are a very underestimated perch bait in my opinion and I should imagine that the perch were preoccupied with maggots at the weekends and could not be caught on anything else at those times. It was impossible for me to fish during the week at the time, so I reluctantly decided to draw stumps and move to another water.

Although not one perch was caught from the police diver's pit, its story has been included as it is often possible to learn as much about perch by not catching them as you can by actually putting them on the bank.

River Perching

by John Bailey

Were it not for signs of real hope, I would not want to write this chapter, depending as it does on the successes of the 1960s and 1970s. The hope of which I speak seems real enough: there appear to be signs of revival in my home rivers, the Wensum and the Bure; the Suffolk Stour is said to be producing fish, as is the Trent and its tributaries. There are perch showing in the Wye and the Tay and all manner of rivers between. I do believe that there are as yet undetected pockets of perch the whole country over and that as the 1990s progress, catchment after catchment will be uncovered.

THE PERCH SHOALS

This, anyway, is my hope, for river perching can be quite unbelievably good. It can be vivid, exciting, and if a shoal is contacted, exceptionally prolific. The discovery of a river perch shoal has at times provided me with some of my greatest angling moments. My fondest memory is of the shoal that roamed the Upper Wensum in the vicinity of the Anglo-Saxon cathedral in those reaches.

I discovered this group of fish by absolute chance. At the time, in 1972, I was working on the land and a ready supply of worms was available to me, ready delivered along the potato riddle that I manned. All summer long, these had been my bait for evenings on the river and I recollect they had caught me just about everything: big old jacks, roach, a good bream or two, weir pool tench, odd great brown trout, eels and at last perch. And how!

It was about 7.00 p.m. on a lovely July evening and I hurried on to the plain with two hours' bliss before me. My target was a newly-found swim where an alder collapsed into a slow, deep, eddying bend. As I tackled up to trot a worm the fifteen yards into the area, I expected roach and when I hit the first bite I did not expect to be disappointed. The fish was obviously a good one and had jagged in the good old roach way. A yard from the net, its spiked dorsal held high, I realised my mistake. Desperately excited I fumbled for my scales: they registered 2lb 2oz.

A pristine little river perch taken by trotting the stream.

By nine, my net held seven such fish. The largest weighed 2lb 7oz, and the smallest was 2lb 1oz. I laid them in the rushes and simply marvelled. Their size, their colours, their glaring defiance seemed as splendid as anything I had seen before. I had no camera in those far off days, but I have no regrets – the image of that night is with me still: it will remain forever.

Even as the summer began to give way to autumn, I continued to hunt that shoal. On my next session I landed four fish and that moved them. Thereafter I always needed to rove until I located them once more and would generally take two or three fish until they wised up to my return and ghosted away into the weed and untroubled water.

Sometimes I caught fish I recognised, but more often not. My biggest fish by October was 2lb 10oz and my smallest was 1lb 14oz and I began to guess that there were around twenty fish in the shoal. That was until one Sunday when the sun was bright and the river still ran crystal, summer clear. I found the shoal at last where the river entered a long and thick copse about midway between the mills. I took the first fish – 2lb 4oz, I remember – admired it and prepared to trot down for the second.

For some reason I stopped fishing and decided to try and spy on the shoal for the very first time. I noticed a good climbing tree close to where my fish had taken the bait and made off through the undergrowth towards it. Never have I climbed a tree more cautiously: every crack of branch made me

A vivid river fish.

shudder and pray the perch would not be moved. They were not. I reached a look-out spot ten yards up and looked down on the river. My eyes scanned the weedbeds and focused on the stones of the bed. I began to see fish. Perch, perch and more perch. I gasped. I began to try to count them. They would not stay still and milled around and around till I was confused. A total though began to emerge: I was seeing a shoal of between 100 and 150 perch. What I had caught had not even scratched the surface of this immense fleet of fish. I had just never guessed the magnitude of the shoal I had pecked away at. Those were indeed great days!

As I came to know more about the river and its fishermen, I heard more about equally huge perch shoals of the past. Along all the Wensum these bands of fish patrolled, dominating an area, grazing on all the food supply there and then moving away to new feeding grounds. History had it that some of the shoals had held even bigger fish than my cathedral specimens. Catches of four and five 3lb perch were not uncommon. Once, two four-pounders had fallen in consecutive casts. Would that I had been present to witness that!

How long a perch shoal tarried in an area depended on the food available and whether or not angling pressure was exerted. In my experience, the capture of five to ten fish would be enough to move the shoal. On a couple of occasions I surveyed the area vacated as carefully as I could: there seemed to be very little left in the way of food. All free-swimming small fish like minnows were gone: the bed was vacuumed completely clean of gudgeon and loach. I could find no crayfish under any of the stones and even the weeds seemed to have been stripped of their water snails. There were absolutely no dace or smaller roach in either area. In both places an area of about 50–70 yards appeared to be quite stripped of food: I believed then that a couple of hundredweight of marauding perch made an effective scavenging machine, and woe betide the little fish of a swim so chosen.

Obviously this shoaling characteristic is not confined to the Wensum. Certainly, during the mid 1970s to early 1980s the Upper River Yare also had at least one such body of perch. They did not grow as large as their Wensum counterparts – the average size was 1lb 8oz to 1lb 10oz – nor was the group nearly so numerous. From slightly obscured sightings I guessed their number to be between 30 and 50. Yet, even without the glamour of the 'Cathedral Shoal', these Yare fish did give me a great deal of pleasure. The river was so lightly curving and feature-full I often found the perch literally under the rod tip and then great battles would ensue. Yare fish were far more cautious than Wensum perch, perhaps because competition for food was less, and finer tackle was a must. Frequently, all they would reach to were single maggots on a size 18 to 1.7lb line and then, boy, watch out!

THE LONERS

Besides these remarkable shoals of perch, all the Norfolk rivers held what we called 'loners'. These fish lived alone, very rarely roamed from their own particular lair and frequently grew very large. They were also very difficult to catch and my regret is that I only ever did land what was probably the smallest loner that I saw. He weighed 2lb 4oz! One fish that lived under Guist road bridge on the high upper Wensum could not have been less than 4¼–4¾lb. I saw him often; so did my friends; so did many other anglers as far as I know, but I doubt if he ever came out of that gloomy hole under the concrete arch. He would follow spinners gladly, he would be happy to bristle at a livebait, and he would oblige by glaring at a worm but, of course, he would decline to put any of these in his mouth.

'My' Guist fish was typical in so many ways: his size, his caution, his affection over two years for one particular lie all made him the representative of his breed. Why he had become a loner I did not know. Perhaps his

The typical lair of a 'loner'.

shoal died around him and he was the last survivor. That would explain his size, and it would add a touch of glamour to the old boy that I liked – but it might well not be the truth. Quite probably he preferred the solitary life and found that as a single fish he need not travel, but that food would continue to come to him.

I do know that several other loners did exist in those perfect perch days of the 1960s and 1970s. There were several in mill pools here and there, haunting oxbows or lying low under bridges. One big perch lived for three seasons in an underwater cabbage patch and another in a fallen tree that straddled a backwater. All I can say is that big loners looked hard for cover, rarely went far from refuge and were notoriously hard to catch. Shoal perch were easy, then twitchy and then moved on; loners were not, of course, impossible, although I might have believed them so. I realise now that my techniques for them then were not sophisticated enough – nor perhaps was my dedication sufficient. The 'easy' shoal fish always beckoned and, keen to see a bend in my rod, I was always eager to follow them. Now I would spend a week after that one Guist Bridge fish alone – perhaps even a month or a whole season! Why not, if that would put him on the bankside?

A perch-inhabited River Bure mill pool.

THE METHODS

What I did then, a number of years back, was pretty sound I think, and I would not change my approach a great deal – for the shoal fish at least. For them the trick was to long trot with a good, big, lively lobworm under a float that hardly needed to be shotted exactly. Indeed, the weight of the worm could nearly make up for a swanshot! It was important that the worm should swim about two feet off the bottom. In normal 4–6 feet deep runs, the perch tended to be at that level and also the worst of the bottom weed could be avoided.

A twelve-foot rod, and a fixed spool or centre pin reel loaded with 4lb line are perfect for this job. Bites tend to be suicidally positive, so much so that it is vital to hit them as soon as long-distance fishing on a lively stream will permit. The beauty of long trotting is that a man can fish out 30–40 yards in 5 casts, each one following a different line downriver, and then move on. It is roaming work, but the shoals will be found eventually and a few good fish will be the result. It has always paid me to travel light. The trout man's approach is best, with a shoulder bag of bait, hooks and floats, a small net from the strap and the rod itself. That way, you can go for miles, in heat even, when the nettles and flies are bad.

Perch like these make river fishing a joy.

A favourite bait – fresh-killed fish.

It was my mistake to fish in much the same way for the loners. I always sat upstream and trotted delectable baits to them – and always, sadly, past them. Now I would tether baits in their area and give them far, far more time to inspect and make a decision.

By far my favourite bait would be fresh-killed fish – a small roach or gudgeon. I would let it travel along the bed into the perch's vision and then stop it a foot or so out. I would sit over it then, if necessary for several hours and certainly well into darkness, when I feel these very large fish come out to scavenge.

Larger Pits

by John Bailey

Big pits have always exercised great hold over the perch fisher's imagination – or at the very least since the 1950s and Richard Walker's great catches of the species from Arlesey. He made history there in several ways: the size of fish caught was astronomical; the distances he cast were so great he had to design a new lead to get him there – the Arlesey bomb; the depths he fished at were also greater than almost anyone had contemplated before. Before those heady days were at an end, he had stimulated a great deal of discussion into his methods, and, of course, desire in others to emulate him.

My own chance to fish for big pit perch occurred primarily in the years 1973–75, and over three seasons I did a great deal of work along the Walker lines. I must say that my results were only average. I now believe that the pit I fished was not a prolific one but, also, I realise that my techniques then were so limited I could never have hoped to do really well. Yet I enjoyed myself with fish in the 1½–2½lb range and once came into close contact with a fish that onlookers could only describe as near a record . . .

The day had been slow, ever since the October dawn. At around midday, however, there were a few twitches on the bobbin and one materialised into a steady run. Even at 40 yards, in deep water, I realised it to be a smaller fish and began to pump it rapidly towards the bank.

At the cry of 'fish', a couple of friends had gathered on the high bank behind to watch the action. It was they who first saw the hooked fish in the clear water. They called down their guess of a pound for it and prepared to walk back to their own rods. Something, though, caught Billy's eye and he looked on for another ten seconds in silence. Then he spoke, 'Oh, my God! What on earth is that? Look there. Following.' Joe had joined him. 'Ease on the fish, John,' he said. 'Let it run. Give it some slack. Now! Now! Do you hear?' I did as I was told – though much astonished by everything up above. Line began to trickle through my fingers slowly at first, then it literally belted off in a blur. A loop caught on the bail arm. The rod thwacked over and everything, little perch and all, was gone. It was a huge perch they had seen following the hooked fish up from the depths. They would not be drawn absolutely on size, but it had been big enough to seize the pounder

An optimistic Roger Miller at dawn.

and it had utterly overshadowed a 3lb 7oz fish Billy had seen two days earlier. He put it at five, perhaps more, but we never saw anything like it in the pit again. That was the end of it, though I still feel it thump now and again down my rod arm when the perching is slow and I let my memory roam.

That fish, lost though it was, represented the highlight of many dour days. I realise now that my baits were not the best I could have used. Then I tried small dace almost exclusively; now I understand that these fragile fish could not stand the impact of a long cast or the rapid pressure changes down to over 25 feet. Now, if I could get them, I would be happier with something tougher. Gudgeon, loach or bullheads would do the job, or even better 3–4in mirror or common carp, if I could bear to use them. I do believe I also overlooked the advantages of simple lobworms – after all, they had done Walker well enough at Arlesey.

My end rig was not all it could have been either. I employed the simple moving lead, and the majority of the time I suspect the bait lay dead or dying

109

on the bottom. Sometimes it would be picked up by an actively feeding perch but it would not attract mid-water prowlers or incite any attack from fish disinclined to feed. Much of the time, the barely moving dace must have lain half-hidden at least in bottom silt and weed. Today, I would use, indeed *do* use, much more the sunken float paternoster which keeps a bait up and active in the attack zone a foot or more off the pit's bed.

When I did register a run, all too often it was dropped quickly. This I believe happened because the perch was not really interested in the dace but acting more out of curiosity. Also the running lead set-up could easily foul on bottom debris and at this first sign of resistance the gingerly held bait would be discarded. When a run did develop nicely and I let it go to be sure of a good contact, too often the bait was down deep and I injured a good fish. I did not then, as I do now, have the knowledge or nerve to vary the hooks and hooking arrangements. Small trebles and mini snap tackles would have served me much better 15 years ago than the simple single size 8 through the lip.

I also made errors of swim choice. In the summer and early autumn, it was often possible to see perch hunting in the shallower bays at dawn. They

Bringing perch up from deep is a problem.

A happy perch fisher with regard to his quarry.

were fine fish and at times one would fall to a surface fished bait or even plug. However, once winter had set in, I continued to waste precious time on these 4–6 feet areas. I disregarded Walker's pioneering work and only turned to the deepest holes slowly and never with full belief or commitment.

And to be truthful, I never spent nearly enough time charting the water. I believed I knew where deep water was and made a few cursory soundings, but it was all very hit-or-miss, as my results proved. In this last season, I took a Humminbird recorder on to the same pit, and its instant read-outs made embarrassing commentaries on a water I thought I knew. I did *not* know the deepest areas. I did *not* know the depths of the areas I had fished for so long. I had no idea of the gullies or ledges or other features at my disposal. Probably even worse then, I disregarded the areas of discarded machinery. The Humminbird showed up excellent cover over a wide and deep area, and I have no doubt that in the perch heyday a great deal of the fish centred around those swims.

I did, however, realise by 1974 that some of the perch I was catching from 20 feet plus were not surviving. Walker himself had hinted at the problems of rapid depth changes and I was aware that there was, indeed *is*, a problem.

Quite frequently, perhaps one fish in every three landed, a perch would float up after being returned. I believed then, and now, that virtually all of them died. To keep on fishing proved to be a tussle as much with the conscience as with the perch, and in a very perverse way, I was relieved when the disease stepped in and solved the dilemma for me.

Today I do not know what I would do if I contacted perch again in the 20-foot plus zone. Nigel Witham makes it quite clear in his piece on reservoir fishing that it is probably only by boat fishing that a perch can be brought up, weighed and released quickly enough not to cause it distress or even death. To play a fish from the bank necessarily takes more time and tires the fish more greatly. When released, it is more difficult for the suffering perch to return to its former depth level, unless there is very deep water close in to the bank.

All these considerations make me reflect that even if I knew of big pit perch – and I have several suspicions of good fish that do exist – I would not go after them between November and March. I would have to restrict myself to the summer and autumn when I could pursue them in the high water zones and not wreak havoc by inflicting on them pressure changes with which they cannot cope.

You might think me over-sensitive; I do not agree. I prefer to think that perch fishers are not out of the modern day fish-at-any-cost specialist mould. I would hope that perch fishers show more consideration for their quarry – and for each other – than is usual in the frenetic worlds of carp and pike fishing. This does not mean to say that I regard perch fishing as a backwater in any way. Far from it: I simply mean that in this arena perhaps sanity can still survive.

A Perch Fisher's Weekend

by Roger Miller

The delayed winter I had feared for months had at last arrived and the perch rods were sadly redundant for the first time since the late summer. I had several perch waters I was eager to try before closing day but with the ides of March seeing the only serious snow and ice of the winter I was frantic that time would run out on me. Rumours were all I had to go on but they had hinted at an authenticity of sorts which troubled me – I simply had to investigate! A serious stint on each water was out of the question as only days of the season remained, but the month was at least mellowing out into a mildness of sorts that I had so craved. I was blessed with a three-day break off work that would see me on the perch trail once again. The recurring theme of all the waters I had set my heart on was the three-pounders that they reputedly contained – these waters, all as different as they could possibly be, awaited my perch fishing ambitions.

A tiny, clear, weedy little pool that bathes in the neon lighting of the

Perch fishing in a wooded pool.

adjacent town at dusk was, I have to confess, not the most enchanting of my choices; with litter everywhere, dogs and their kindred by-products and also the hum of the traffic, I immediately wished myself back on one of my beloved estate lakes.

The orangeness of the neon glow suddenly snuffed itself out as I wandered down to the water at first light. The beauty of the pool was not in its surroundings, quite obviously, but in the numerous perch-holding features it contained. With a day at the most to spend there I was forced to read the water as best I could and stand by my judgement.

In the centre of the pool were a number of withered stumps, bleached white by the winter months. With a depth of ten feet around them I was convinced that most of the perch spent a great deal of time in this sanctuary, only coming out to feed. The matted stems of the labyrinth also looked a probable spawning site.

What really motivated me about the pond was not only the three-pounder that just might have been caught, but friends had actually taken eels to 5lb 14oz a couple of years before. These had evidently grown big on the masses of tiny roach that flourished in the pit – perhaps the perch had in fact grown big alongside the eels.

I cast two gudgeon livebaits on sunken float paternostered rigs to land within a foot or so of the sizeable area of entangled roots and stems. The wind tore down the pit and the line was pulled out of the Gardner run-clips at least a dozen times. Five hours of this farce saw me reel both baits in with a mind to find a more sheltered spot. The tails of both baits had been chomped! Resistance! If big perch feel anything as they take they drop the bait immediately, and they do not come back. The clip's settings were as light as the wind would allow – the ways of big perch can be quite maddening at times!

Somewhat disgruntled I drove 40 miles to what can only be described as a rich man's folly – a series of tiny lakes set in well-keepered woodland, dotted white with thousands of early snowdrops and linked by magnificent arched stone bridges.

The schoolboy had described the perch predictably as 'whoppers' and with the biggest at 3½lb he said I simply could not fail. The problem for the experienced perch fisher when told of these things is often whether to actually believe them. It is far better to accept that perch have been caught and forget everything you have been told about their size. I work on the assumption that where there are small perch there are often big ones. Fishing for fish that may not exist can sometimes be a help and sometimes a hindrance.

This estate lake outfall pool is the classic perch haunt.

I fished that small system hard all afternoon till dusk, but the tireless little gudgeon had covered every likely-looking perch lair to no avail. Two 'hot' tips followed up in a single day and nothing to show, such is a perch fisher's life!

The following dawn the alarm failed to wake me. At the new pit I was confronted with a frantic John Bailey tearing at the soil in a vain attempt to gather together a few worms to cast to the striking perch that were everywhere! I was the bait man and John had left home assuming that I would be on time with livebaits aplenty. He was wrong!

Free-roving livebaits were soon bobbing around in somewhat precarious circumstances, and within minutes both of us were holding two-pounders – admiring them, photographing them, drooling over them, drinking in the quintessential pleasure of hitting a new water just right. That new water revealed its secrets to us that morning like no other had done for months.

Suddenly John was in to a sullen old fish that stayed deep. Anxious glances to one another were exchanged as a powerful run towards the alder roots took line off the clutch. Anxious glances became beaming smiles as I scooped out a real bull of a perch weighing 2lb 12oz. The pit at that point suddenly died on us, as perch waters tend to do. A perch fisher can either slog on at the hard waters for a possible leviathan or he can keep moving waters until he strikes gold as we did on that day.

The new pit revealed its secrets at dawn.

The intimacy of small water perch fishing is encapsulated in this shot.

The following and final dawn of the season saw me pushing through the narrow woodland path to an estate lake that I had only seen once before, on the map during my journey to it that day! In my pocket was the letter I had received from the owner confirming that a perch of 3lb 1½oz had actually been witnessed by him earlier in the season. Yes, perch abounded and no, nobody every fished for them, his letter continued.

By 11a.m. I had had twenty runs with a dozen fish between two and four pounds, all taken on long range, sunken float paternostered livebaits. The only minor irritant was the fact that all the fish I had landed were jack pike! The keeper called to see me and mention of my abortive perch session caused his hearty laughter to reverberate through the trees and above the roar of the lake's outfall. Two thousand perch had been netted out by the Anglian Water Authority, he told me, and had been subsequently stocked in a distant broad of massive proportions. What was worse was the fact that they had been replaced with 250 3lb carp – total sacrilege in my prejudiced opinion!

An estate lake perch makes a final bid for freedom.

The beauty of perch fishing on estate lakes is obvious.

He went on to say that one or two perch came and 'slept' hard against the dam wall every night and if I stayed till dusk he would show me where. I fished on but the thought of the two thousand perch not there anymore managed to rather dent my enthusiasm! I had, however, heard of the broad in question suddenly throwing up odd perch again, never dreaming them to be one of the same. So much for life's little ironies.

With a fair proportion of the day left I drew stumps in the assumption that the three-pounder was probably the only big perch left in the lake and that the supply of jack pike was probably inexhaustible. On I motored to yet another fabled perch pool. The petrol attendant in the village, who wanted to chat all day it seemed, noticed my muddied apparel and sundry gear and asked me how I had fared.

'Perch you want? Do you know about the two ponds on the next estate? I've caught lots from them in my time, up to last year – the family keeps me in now but I'm sure they are still there. Try the hall moat at X,' were his final words as I roared off the forecourt in pursuit of yet another perch trail.

119

And so I continue. Like the game of battleships it's miss, miss, miss, but eventually you hit it right and all the dead-ends, the unfounded rumours, the schoolboy fantasies, the bare-faced lies, the frustrating streams of jack pike and of course the Anglian Water Authority are all forgotten when the surface is broken by a fry-chasing perch which the inimitable Francis Francis described as 'with all sail set, their fins extended, their spines erect, as if they meant to devour it without hesitation!'

Rigs and Tackles

by Roger Miller

Whilst certain references to rigs and tackles have been made in the preceding chapters of assorted days and ways it is perhaps necessary to devote this chapter to enlarging and illustrating the proven rigs that served us so well. Certain problems, as you will have read, were encountered and the resulting tackles went some way towards solving them. Undoubtedly, other, different problems can and will be solved by different ideas but we feel that our methods served all our waters well. What follows is a series of diagrams with explanations of their application and use.

HOOKS

Partridge Size 14 Treble

This is by far the strongest treble hook on the market for its size – and so it should be as it was designed for the mighty sea trout. No perch will ever straighten or break one of these. Its small size is a real boon because, in our experience, big, resistance-conscious perch are not lovers of large, heavier pike trebles. Drennan Specimen Plus line (4lb b.s.) complements the hook perfectly. There is no need to go above this gauge of line for perch – indeed, double-figure pike have been accidentally hooked but successfully landed quite comfortably with it. I would, however, suggest that it is changed regularly as sustained perch fishing wears the line faster than fishing for the cyprinids with conventional methods.

A simple grinner knot tied through the eye of the hook is preferable to wrapping the line around the shank and under the bend as is the custom with wire. Nylon is exposed to excessive wear if the treble is tied in this fashion and as we all know the line will decide to part just when an enormous perch is at the net.

121

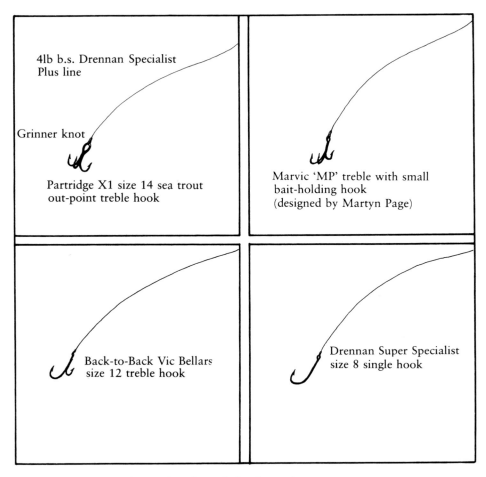

The four proven hooks used exclusively by the authors.

Marvic MP Treble

Another very useful treble to the perch fisher is a recent development by the Marvic Fishing Tackle Company. The hook incorporates a substantially smaller bait-holding point that allows the treble to be struck free of the bait far easier. These trebles are made as small as 12 and work most effectively.

Back-to-Back Double

Fellow perch fisher Vic Bellars designed the Back-to-Back double hook many years ago now and has been using them ever since. Exceptionally strong with the small bait-holder hook soldered on to the shank, the single, larger hook stands proud and unhindered upon striking.

Drennan Super Specialist

Peter Drennan's Super Specialist single hook is the best on the market and is used by us when fishing lobworms at all times. Every fish that swims loves lobworms, from great carp to giant eels. If one of these big powerful fish is hooked, and they occasionally are by perch anglers, this hook will be perfectly adequate.

It has to be said that, for live and deadbaiting, trebles beat singles as they do with pike simply because you have three times the chance of achieving a good hook hold in the bony tissue of the predator's mouth. A change to a single is usually implemented when the perch are feeding hard, as a treble too deep inside a perch will easily kill it. A single carefully removed at such times increases the chance of the fish's survival should this unfortunate deep hooking occur.

HOOKING ARRANGEMENTS

Livebait

Although perch tend to cripple their prey by attacking its tail they often consume them head-first. A lip-hooked bait can be struck quickly once the run is under way. The size of the shot is naturally applicable to the size of the float.

Lip-hooked bait fish of the size used for perch tend to swim with their heads pointing downwards not least because of the weight of the treble. A bait hook in the dorsal root is a useful progression when perch become suspicious of bait acting in the lip-hooked manner. The swimming movement of the bait is thus altered, increasing the effectiveness of free-roving livebaits. Timing the strike is the same. Baits hooked in this way, however, do tend to expire rather quicker than lip-hooked ones for obvious reasons.

Deadbait

Lip-hooked deadbaits for perch accommodate an instant strike facility under normal circumstances. It has been our experience that for reasons best known to themselves some perch consume the deadbait deep on the spot or within a couple of seconds of the take. This means that when a deadbait is lip-hooked under these circumstances deep hooking can occur. By simply placing the hook in the tail of the bait the chances of deep hooking a perch are minimalised as the hook is now out of the danger area.

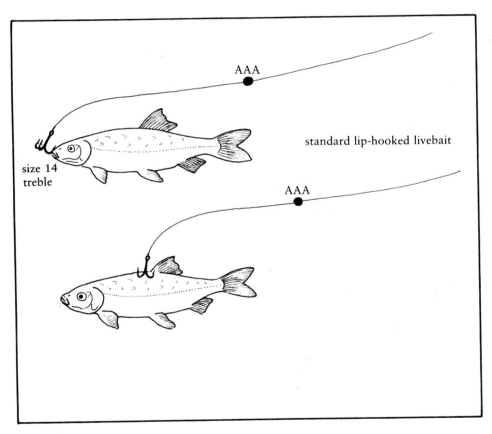

Livebait hooking arrangements.

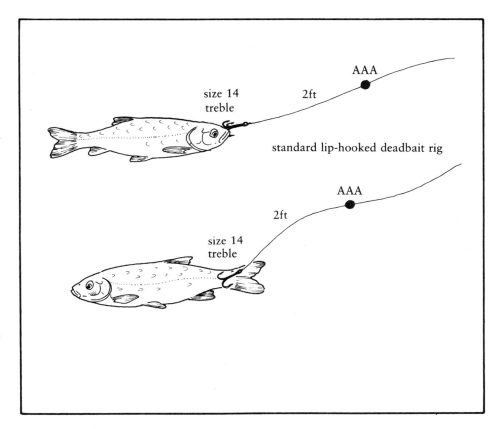

size 14
treble

2ft

AAA

standard lip-hooked deadbait rig

AAA

2ft

size 14
treble

Deadbait hooking arrangements.

Tail-First Attack

Chapter 7 identifies the problems encountered that eventually produced this rig. Essentially it was a rig to overcome the problem of using baby carp as livebaits. Whilst the perch absolutely adored them a real wrestling match no doubt occurred underwater whenever one of the carp was taken! Transmitted to the float it was impossible to time the strike effectively, for the carp would be towed around with a number of stops and starts in between.

A hook is difficult to strike free of the tail root of a carp and the flier treble seemed the best idea. The 1lb b.s. line tied to the tail allowed the free treble to be struck home. The carp's natural swimming motion was not as restricted as it is with the treble inserted in the actual tail.

Tying the treble to the tail of the lively little carp is difficult but the best way is probably to cover the fish in a wet cloth and use pre-tied loops of line, slipping the loop over the tail and pulling it tight. The old fashioned granny knot served me well here. A single hook is all that is needed in the lip of the fish to take the pressure of the tied-on treble. It is also important to keep the length of line between the single and the treble flier slack for the very good reason that the bait appears more natural that way.

Although the rig appears rather elaborate, in its defence I have to say it did solve a particular problem quite well.

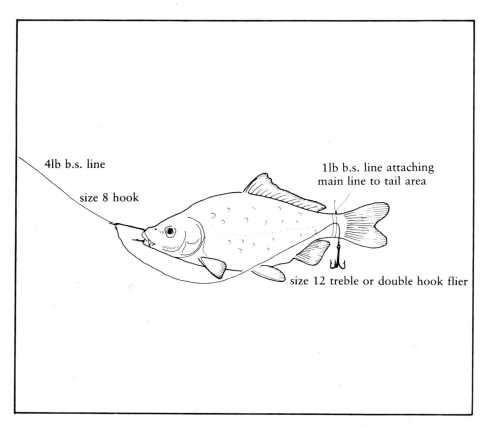

4lb b.s. line

size 8 hook

1lb b.s. line attaching
main line to tail area

size 12 treble or double hook flier

Hooking arrangement for perch attacking bait tail first and not turning it — often a carp livebait.

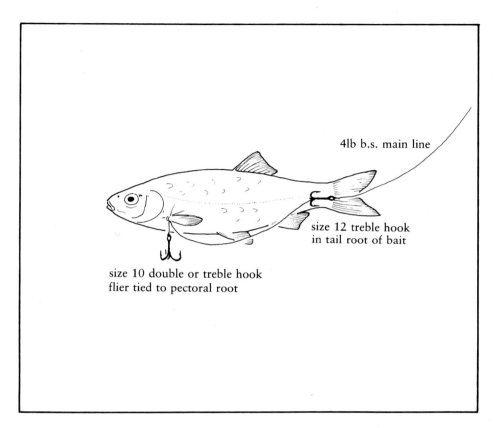

Hooking arrangement for perch attacking bait head first and not swallowing bait — often a dace.

Head-First Attack

Again, Chapter 7 describes the reason for the evolution of this rig, a direct reversal of the original. Unfortunately a hook has to be placed in the tail of the bait, which does incapacitate it to an extent, but the bottom treble, tied to the pectoral root, is deadly, especially when the perch sit with the bait in the manner previously described, or when they wish to consume the bait on the spot.

I later altered this rig by placing the uptrace treble in the dorsal root of the bait. I do, however, suspect that the trebles are rather too close to each other, thereby causing bait ejections. It is early days at the moment and I have no way of proving it. A single hook placed in the tail root of the bait does, I have found, encourage a more free-swimming livebait.

FREE-ROVING LIVEBAIT RIGS

Drennan Loafer Float

After certain trials and errors I found the best float to present free-roving livebaits is a 3 SSG black balsa Drennan loafer float fished double rubber using only 2 SSG shot as weight. I tend to lightly shot the float for a number of reasons, not least because the weight of the bait has to be taken into account. It is important to have the float sitting high in the water to prevent the bait taking it under – a most frustrating experience! Also, the extra length of the float above the surface tells the angler much about what is going on below. The more normal practice of shotting the float down to its tip does not tell him so much whilst livebaiting is being practised. When the float is lightly shotted the loafer float is often seen to tilt one way or the other, or even lay flat, as the livebait attempts to escape an attacking perch. It is also difficult to discern on occasions whether a float is moving one way or the other, especially in wind; the tilting of the float is often a clue in these conditions as to what is going on below.

I prefer the black painted floats as opposed to the crystal (clear plastic) variety as the crystals have a tendency to disappear at low light – the prime time for feeding perch of course! The black of the balsa loafers is more visible on these occasions just as it is in bright weather where the sunlight tends to obscure the clear plastic.

Although Peter Drennan originally designed the loafer float for trotting, it does, however, allow the livebait a greater freedom of movement than the alternative poly-ball float mentioned in the next section. Due to the slimness of the loafer float less resistance is felt to the striking perch which is probably the best reason of all for using this type of float.

I fish the float double rubber style in order that quick changes of float can be effected should a longer cast be required quickly – for example perch striking further out – or alternatively when perhaps a smaller, lighter rig is required for wary, close-in loners. Heavily greased line is important when fishing free-roving livebaits, so that the bait's movement is not impeded by sinking line and also because a floating line will often show the direction of the take and enable a solid strike to be effected in the opposite direction.

Pilot Float

For all the advantages of the Drennan loafer float the Andy Barker pilot float with its 1in diameter size has very useful applications in perch fishing. I

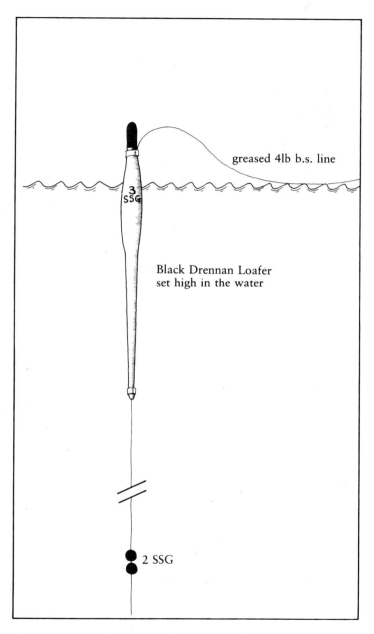

greased 4lb b.s. line

3
SSG

Black Drennan Loafer
set high in the water

2 SSG

*Standard roving livebait rig for perch, with Drennan loafer
float.*

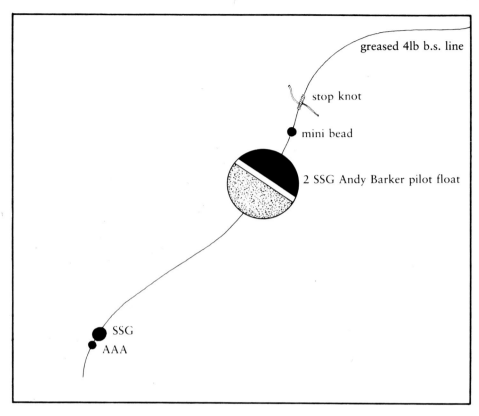

Free-roving livebait float rig.

tend to use these floats during rough weather when wind and rain can make the Loafer float difficult to use, especially when three are in use at once and being constantly blown flat by the wind. Again, I lightly shot the pilot float in order to accommodate the weight of the bait. These floats are unaffected by the wind and remain highly visible. Although an increase of resistance has to be accepted by the perch it does not seem adverse enough to make too much difference.

Medium Middy Floater Controller

The best of both worlds, these tiny little floats are highly visible but very light indeed. They will fish small livebaits yet are still light enough to be used for lobworms. I find them a must when the wind gets up and makes the

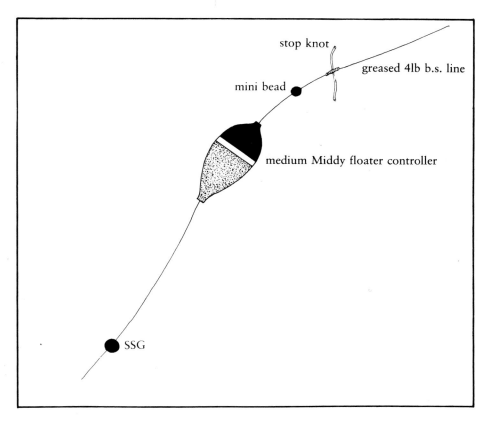

Medium Middy floater controller for close range livebait work.

more conventional floats used for laying-on with worms difficult to spot.

Whenever I use any of the three floats mentioned I use an optonic coupled with an adjustable Gardner Run Clip set at its lightest setting. It is impossible to concentrate fully on three floats all day and anyone who has done any perch fishing will know that one run is often quickly followed by two more to the other rods. Takes can be fast and furious at times and on many occasions the bleeping of the optonic has alerted me to takes when I have been occupied with another fish or whatever. Some perch take livebaits at incredible speeds and to fish without the run clips is stupid in the extreme.

DEADBAIT RIGS

Marvic Mini-Pencil Float

Disappointed that the commercially made self-cocking, cigar-shaped pike floats were not made small enough for use by perch fishermen I was elated to be given one of Marvic's new mini-pencil self-cocking floats, designed for very shy pike but also very good for perch.

These floats are superb for casting a deadbait to a good distance, while they remain very visible whatever the weather. Fished in the conventional slider manner as shown I think it a good idea to place the shot at least 2 feet from the bait in order to lay this length of line on the bottom and out of the

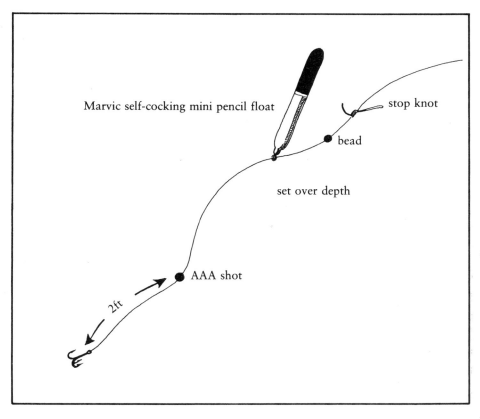

Long range deadbait rig, with Marvic mini-pencil float.

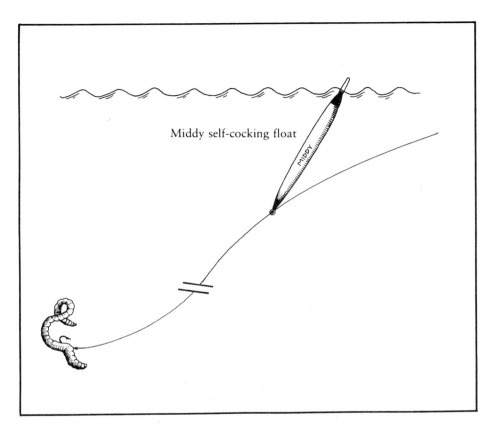

Middy self-cocking float rig.

way of an enquiring perch. I have fished this rig whilst deadbaiting for pike for many years and have total confidence in it.

Middy Self-Cocking Float

The Middy series of self-cocking floats are ideally suited for presenting a weight-free deadbait or lobworm on the bottom. No lead on the line is required at all and a generous cast is still possible due to the natural weight of the bait. It is always a good idea to strive for as little weight as possible in all forms of perch fishing as more runs will undoubtedly result. This rig is one such rig.

OVERHANGING FOLIAGE

Simple Laying-On Rig

On certain occasions a bait needs to be anchored down in a specific spot, for example beneath a tree root or overhanging foliage, as big perch are often loath to move from these sanctuaries. Big lobworms, believe it or not, are perfectly capable of crawling along the bottom and away from the crucial area.

Drag and undertow on some stillwaters can also cause problems that necessitate the standard laying-on rig being employed.

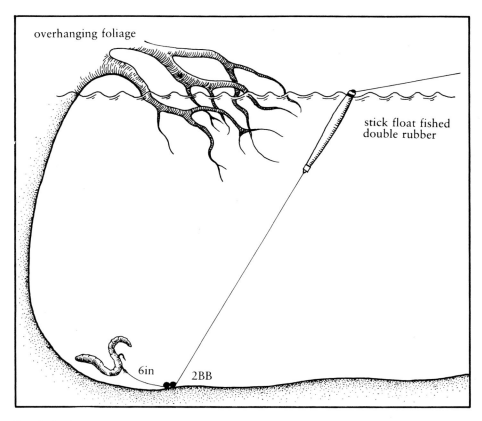

Simple laying on rig.

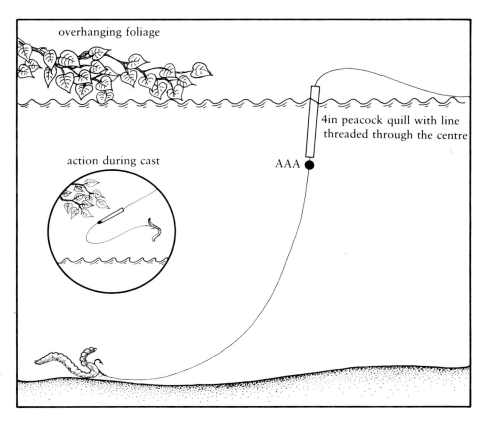

Refined rig for fishing 'overhead cover' swim.

Refined Rig

Precise casting is often the sole difference between a good catch of perch and a blank day. Some big perch seem to actually reside beneath or even in the most inaccessible tangles of roots and branches. Although this rig is hardly new the threading of the line through the peacock quill in the manner shown with the shot nipped directly beneath it does work. Too big a bait, however, can tip the weight balance the other way during the cast. A lob's tail is usually the correct weight in order that the action as displayed in the inset is achieved.

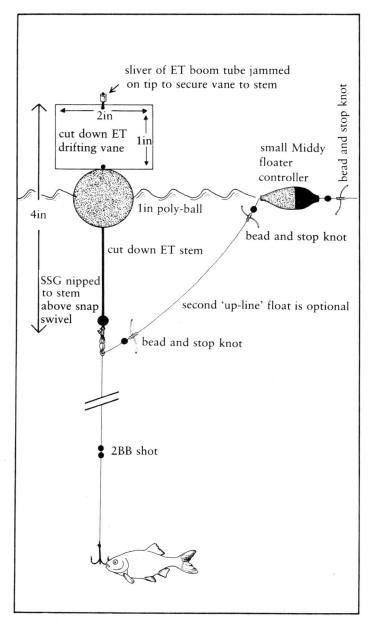

sliver of ET boom tube jammed
on tip to secure vane to stem

2in

cut down ET
drifting vane

1in

small Middy
floater
controller

bead and stop knot

1in poly-ball

4in

cut down ET stem

bead and stop knot

SSG nipped
to stem
above snap
swivel

second 'up-line' float is optional

bead and stop knot

2BB shot

*Scaled down version of Dave Batten's improved ET drifting rig
for pike.*

136

ET DRIFTING RIG

Simple to make, this drifting rig in its miniature version is a tremendous item of tackle for the perch fisher. The larger floats used by pike fishermen require little real alteration but I ensure that the horizontal arc of the drifting vane of the float bought from the shop is turned on its side, rendering the arc of the 'sail' vertical. A more effective vane is therefore created, which is then cut down to the size displayed in the diagram. The length of the stem is only approximate and every float for some reason requires different specifications. I suggest that should you decide to try constructing one of these mini drifters, you start with a stem of about 6in, with the drifting vane, poly ball and SSG in place, and then begin to snip off tiny lengths until the float sits in the water as shown.

Dave Batten's original drifting rig incorporated the tiny float up the line as shown in the diagram. This was an improvement on the small eye pushed on to the top of the stem of Eddie Turner's commercially bought float. Dave's idea eliminates the need to strike out the eye from the top of the float and a much more direct strike is affordable. I have found, however, that when a moderate ripple is on the water the mini drifter can be fished bottom end only in basic waggler fashion. A heavily greased line is again essential for this form of fishing.

For further reading on the practicalities of drift fishing *Pike — The Predator Becomes the Prey* (Crowood) is recommended.

SUNKEN FLOAT RUNNING PATERNOSTERS

Marvic Submerged Float

This sunken float running paternoster rig, although unsafe for pike due to their ability to bite through nylon line, is a deadly one for perch. The sunken float marketed by Marvic was virtually designed for this very purpose by Vic Bellars. Livebaits are presented clear of the bottom at a pre-determined depth depending on the hook length, and with the use of 8lb b.s. line between float and bomb it is virtually tangle free.

This rig is fished under tension straight through to the run clip and the greater the tension, the nearer the bottom the bait is pulled. As tension is released the float draws the bait nearer to the surface. Due to the lightness of the float only a very light amount of tension needs to be exerted in order to fish the rig efficiently. The reason for fishing the rig under tension is to prevent the bait swimming around the paternoster link. Livebaits tend to try and reach the sanctuary of the bottom when they are fished in this manner but the float obviously prevents this, leaving the bait very attractive to a marauding perch.

Drennan No. 2 Subfloat

On occasions, the undertow present has a habit of exerting pressure on the line that in turn pushes the small float over and therefore deeper, thus affording the bait its prized sanctuary on the bottom. An increase in the buoyancy of the sunken float is required at such times and some of the

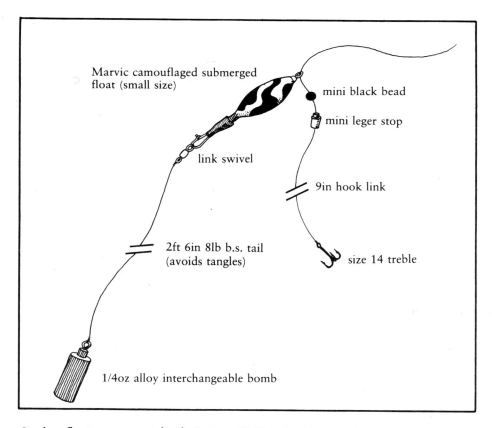

Sunken float paternoster livebait rig, with Marvic submerged float.

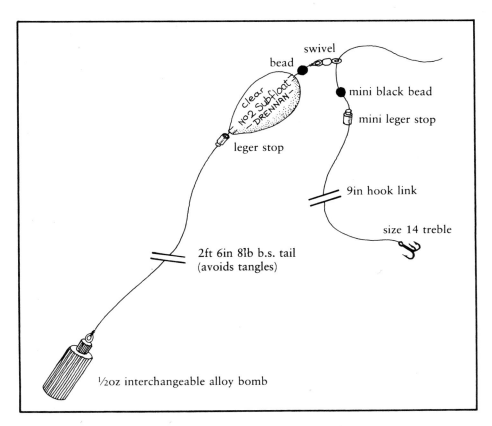

Sunken float paternoster livebait rig fished under tension for long range work or when undertow drags the above rig out of position.

sunken floats marketed for pike anglers by Peter Drennan are ideal. In particular the Drennan No. 2 Clear Subfloat, which is fished in exactly the same manner as the smaller Marvic model, overcomes such problems.

Bite Indication

Fishing sunken float rigs for perch requires very sensitive bite indication so that an attacking perch feels a minimum amount of resistance. For me the most effective method is the use of an adjustable Gardner Run Clip attached to the rear bank stick as shown. The run clip is set so that the slightest pull will release the line from the clip, rendering the whole set-up resistance free.

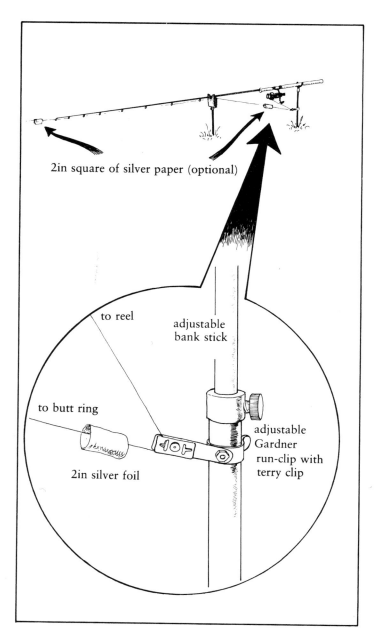

2in square of silver paper (optional)

to reel

adjustable
bank stick

to butt ring

adjustable
Gardner
run-clip with
terry clip

2in silver foil

Sunken float running paternoster bite indication set-up.

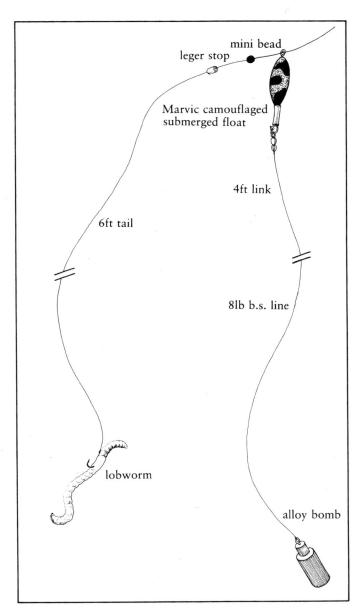

A neutrally weighted lobworm with injected air on a very
long tail can induce perch to take as bait sinks slowly
through the water. A long tail also reduces resistance during
early stages of take to virtually nil.

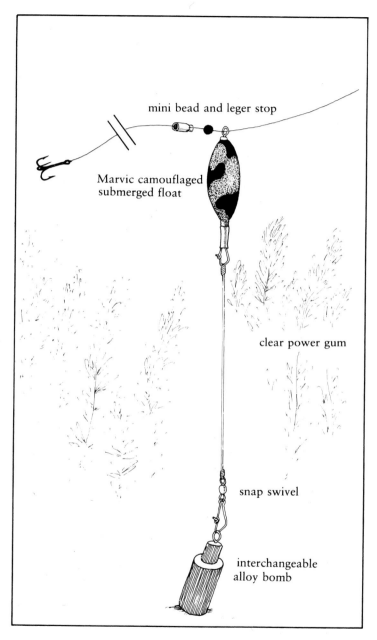

Marvic camouflaged submerged float for fishing worm or deadbait on or over weed following the injection of air.

It goes without saying that the bale arm of the reel is left open, something I always keep forgetting!

Together with an optonic bite indicator I like to use a 2in square of silver paper made into a cylindrical shape and placed between reel and butt ring, as I want a visual form of bite indication for certain reasons. I like to fish rods at 15-yard intervals along the bank and should I fail to hear the optonic or it packs up, the piece of silver paper jamming into the butt ring leaves me in no doubt that a run is taking place! The cylindrical shape allows a free passage of line thus maintaining the resistance free set-up.

The optonic is invaluable, especially when sport is slow and attention begins to wander away from the rods, the bleeping often brings me back to reality! More seriously, the more ways you have of indicating perch runs the better, not least because of their capricious habit of not always hanging on to a bait as long as the angler would like.

Another preference of mine is to nip a small square of silver foil to the line above the tip ring as it is often difficult to discern whether the line is simply laying on the water or being pulled along by a run. Wind tends to peel line off the spool but the small piece of silver paper disappearing down into the depths leaves you in no doubt that a run is taking place. I like this idea also for the reason that the direction of the run can be determined together with its speed, allowing a resulting strike to be in the opposite direction and of the right force.

Fishing Weeded Waters

This rig is essentially a sunken float running paternoster as described previously. It is a favourite rig of mine on waters with excessive weed. The sunken float keeps everything above the potential resistance of the weed thus eliminating the chance of the tackle becoming caught up. A neutrally weighted lobworm with injected air, fished on a very long tail, can often induce a big perch to take as the bait sinks slowly through the water, ideally into a gap in the weeds. The long tail also reduces any possible resistance to virtually nil.

Big wary old perch love to lie in small clearings in the weed and a worm fished in this manner is often irresistible to them!

Fishing on Weed or Mud

Another version of the same rig which is in reality an elaborate link leger, this rig allows baits to be fished *on* weed or soft silty bottoms so long as the

143

lobworm or deadbait is neutrally weighted with injected air. The sunken float, the same as used in the sunken paternoster rig, allows a free passage of line through its eye as it holds the line above the bottom. The length of the link is in accordance with the depth of weed or mud you wish to fish over. Again, as always, resistance-free fishing is the prime objective.

Conclusion of Events

by Roger Miller

Looking back over the mass of perch fishing we have undertaken over the last couple of years, one word is essential – organisation. During our occasional forays into intensive pike fishing we thought we knew what organisation meant. We were wrong.

During pike quests, we have of course had to organise a livebait supply but often we would not bother, relying instead on frozen deadbaits from tackle shops, or even fresh herrings from the local fishmonger. No such luxury exists for the perch fisher. Sea baits are hopeless for perch and both of us would rather leger a sprat for roach than for perch! Indeed, frozen coarse fish deadbaits are nowhere near as effective as a freshly-killed bait. Therefore a supply of healthy live fish is essential, whether you intend to use them live or dead.

We think it important to outline briefly our philosophy on the use of coarse fish as bait, either live or dead. We base our argument on two distinct points, namely, conservation and cruelty. While we dislike the use of live fish for bait we continue to do so because they are necessary on many waters in order to catch big perch. We can just about live with this aspect in our quest for bigger and better perch.

The conservation issue is far more important to us, as many fish from environmentally threatened areas are still being taken as bait, together with fish from the popular match stretches. These acts are unfortunately perpetrated by a minority of pike anglers but happily, an ever-growing proportion of pike men are now buying their livebaits from trout farms. This avoids the persecution of flourishing or perhaps diminishing coarse fish stocks. We would dearly love to be in a position of fishing for perch capable of taking trout livebaits but as yet, unfortunately, we are not! We are therefore faced with the problem of finding a water where a reduction in stock will actually be of benefit to the remainder.

Fortunately a stunted carp pond that was choked with runt fish presented itself. By trapping out quite a few we soothed our guilt at livebaiting with the belief that the rest of the little carp in the pond were grateful to us for relieving them of a few of their brothers and sisters. This is our stand on livebaiting and we ask that all anglers consider the impact of taking coarse fish as bait very carefully before doing so.

Once livebaits have been obtained we believe in taking care of them. Unnecessary deaths through neglect are a waste, as it is very satisfying to return baits to a better environment at the end of the season than from where they originally came, should they not be used on the hook.

The easiest way of keeping fish alive in captivity is to situate a tank at the bottom of the garden, with a hose-pipe continually introducing fresh tap water. A mere trickle is perfect. This is a far more efficient method and considerably cheaper than using an oxygen pump round the clock. Also, the water is kept clean when using the hose-pipe and this allows a sight of your stock so that the odd dead one can be removed. It is imperative that a net be kept over the tank as domestic cats will soon be pawing them out – ours certainly did!

One of Roger Miller's cats caught in the act of snitching livebait.

Organisation even has to run to lobworms, as they have to be dug and kept alive. For perch we feel there is nothing to beat a giant lobworm that has the ability to make a small float tremble when being fished! By far the best places to find such lobworms are the central reservations of street-lit roads or roundabouts. The neon lighting renders the worms relatively hardened to light. Indeed, a torch is not even required in their pursuit. With thunderous traffic passing by most of the time they are not as suspicious as garden worms should you be guilty of a heavy footfall. These urban worms are often incredibly numerous and very easy to obtain.

We apologise to all the traditional 'snitchers' who will undoubtably persist with the dimmed torch/carpet slippers/bowling green approach! We trust you will forgive our flagrant disregard of angling ethics!

We never bothered with the time-honoured 'wormery'. A couple of large livebait buckets, filled with good quality soil and left on the bare concrete of the garage floor proved perfectly adequate. The old books are full of feeding the worms with tea-leaves and shredded newspaper, and just why a worm has to be 'scoured' in moss is a total mystery. We left ours well alone and they did not suffer!

It is, we believe, absolutely necessary when confronted with an unknown perch water to determine the vagaries of its perch by using at least three different baits, namely, lobworm, livebait and freshly-killed deadbait. Actually, to use four is best: one livebait should be free-roving and the other fished on a sunken float paternoster. We can both point to waters where any one of the four proves by far the most effective. When fishing as a partnership all four baits/presentations can be tried at the same time.

The use of wire for perch in our experience suggested that it was about as effective as using no bait at all. Fellow perch fisher, Pete Garvin, has, he tells us, had many of his big perch even when using wire traces. This is most perplexing but as Peter's fish are twice as big as ours perhaps there lies the answer!

Perch are extremely susceptible to even the subtlest changes in light values. Richard Walker, in *Stillwater Angling*, states of Arlesey perch that he never caught one in dull conditions, but bright, cloudless, sunny days were the most productive. Arlesey was of course very deep and his recommendations threw a lot of people as bright, cloudless days in winter became the 'necessary' conditions in which to pursue big perch. This misconception lasted for years. Dick's approach was correct for Arlesey but such conditions are definitely unfavourable on the vast majority of waters we fish that are shallow and often coloured. We feel there is considerable room for research in the relationship between perch behaviour and light.

Some facts are known about the vision of the perch, and we may find some pointers in this complex subject. The eyesight of the perch is highly developed to deal with the diffusion of light underwater, where the colour red dominates. Sea fish that live in water of exceptional clarity, that is water with virtually no algae, perceive blue and green light most readily. Freshwater fish that exist in algae-rich lakes and rivers where the coloration of the water is quite evident have evolved to perceive red most readily. This is the case for all freshwater fish in Great Britain. Perch hunt their prey very much by sight, as well they might, as the pigment of their retinas can absorb light which is further into the infra-red than can any other freshwater species.

The advantage of this ability is that underwater, much of what is seen by fish appears blurred, but the greater a fish's vision into the infra-red is, the sharper the outline of the prey or object becomes. Indeed, the majority of fish possess a blend of pigment in their retinas, which consists of rhodopsin, that absorbs green/blue light, and porphyropsin that absorbs infra-red light. This gives all fish the ability to see more clearly underwater. The

Nocturnal feeding perch that disprove the rule.

Perch that ferreted out deadbaits unerringly.

pigment present in the retina of the perch is pure porphyropsin and this makes the vision of perch totally dedicated to discerning prey fish underwater. In fact the kind of porphyropsin found in perch can penetrate further into the infra-red than any other yet discovered.

It is therefore quite logical for a perch to pursue its prey by sight. It may also go some way to explain why night-feeding perch are very rare.

With all this in mind, many things that anglers have discovered over the years fall neatly into place. The old method of attracting river perch by stirring up the bottom and drawing in small fish like gudgeon is a very good example. Anglers assumed that the perch were attracted to the area because of the congregation of small fish. This was true, in part, but the coloration of the water made it easier for the perch to feed as their superior vision in the cloudy water gave them a killing advantage. Coloured streams flowing into clear lakes are renowned hotspots for perch, for exactly the same reasons as those just described.

My experiences at Greenmire Pool were such that the perch only fed in deeply coloured water, simply because they could see the prey far better than their prey could see them. A final example of the exceptional vision of perch is surely the police diver's experience of them being the most easily frightened of all fish. This was of course because the perch saw him before everything else.

The sense of smell possessed by perch is, however, much underrated. They can ferret out deadbaits just as unerringly as pike, and chopped lobworms draw them in like magnets.

The location of perch is far easier than the location of pike because their presence in a particular area is usually quite logical. As John Nunn's piece on fish recorders states, perch will seek cover under absolutely anything. Perch are, it seems, quite happy to live within the confining branches of a fallen tree, rarely coming out even to feed. Often a bait fished tight against the tree's branches is the only one to be taken. Features that afford the perch the cover they love are numerous. They include boathouses, moored boats, buoys, trout cages, lily beds and roots, bridges – either collapsed or otherwise – trailing branches, fallen trees, weedbeds, tree roots, reeded bays or inlets, islands, stakes, rafts, undercut banks, boulders, sunken boats, any dumped bulky rubbish like fridges and cookers, as already mentioned the stream colouring the lake and of course the time-honoured sunken piles. Perch leave these havens to hunt prey in packs. The larger the perch, the smaller becomes the pack.

We both feel quite justified in stating just how underdeveloped modern perch fishing actually is. The appeal of the perch is enhanced by the

Tiny reedy inlets, with perch swims in all of them.

*A bridge, a bay and a dull day amount to a perch fisher's idea of
a perfect combination.*

151

Lily beds are always a home for perch.

Trust my cat not to prefer 'whiskas'.

knowledge that few anglers are pursuing them and that everything you learn about them is down to hard work. It is amazing to think of all the rapid changes that have taken place in angling since the devastation caused by the disease, but the species is relatively unexploited by these advances.

Very little modern literature is currently available about perch but much of what is written in *Zander* by Barrie Rickards and Neville Fickling is very pertinent to what we have discovered about its cousin the perch. The species are very similar in some ways and we have learned much from transferring the information to our perch-fishing circumstances.

The New Complete Angler by Dr Stephen Downes and Martin Knowelden is worth reading for the perch chapter alone, in particular the detailed section on the vision of perch.

Perch are the most fragile of fish and any amount of undue stress seems to have fatal consequences. This seems to have been taken into consideration by Mother Nature as perch can breed like no other species and subsequent maturity to adulthood is remarkably brisk.

Perch will continue to come and go and we must accept this facet of their natural history philosophically, but to miss them when they suddenly reappear out of the blue is one of angling's greatest mistakes. Of all our messages, there is one which we hope you will take to heart: perch are out there — it just remains to go and get them.

Appendix

THE PERCHFISHERS

by Steve Burke

Elsewhere in this book my fishing partner, Nigel Witham, writes of reservoir perch and in particular the mighty perch of Bewl Bridge Reservoir. Sadly these fish are no more, victims of the infamous perch disease. Even the name is no more, for the 770 acres is now known simply as Bewl Water. But out of death comes life; and out of Bewl came the Perchfishers. For it was here, on supposed pike culls, that three groups of the original members fished, unknown to each other, not for pike but for perch.

It wasn't until the autumn of 1986, two years after the last of the Bewl bonanzas, that we were to meet up and compare notes. And comparing notes is essentially what the Perchfishers is all about, for comparatively little is known about the species from the angling point of view.

Twelve months later the original enthusiasm hadn't waned and it was decided to formalise the structure and open the membership to other experienced perch anglers. There is no qualifying weight for membership, all an applicant has to do is to somehow satisfy the committee that he or she is prepared to contribute to finding out more about perch and perch fishing.

The atmosphere is very friendly, without the ultra-cult tendencies of some of the groups specialising in the more popular species. Perhaps this is because most of us are all-rounders, although for many perch are our favourite fish. Many well-known anglers including contributors to this book are members. Contact has already been established with perch anglers in the USA, Sweden, Holland and Germany and further links with other countries are in the pipeline.

Part of the membership fee covers the cost of support for the Water Charter organised by Friends of the Earth, and affiliation to both the National Association of Specialist Anglers and the Anglers' Co-operative Association. Having said that we urge all anglers to also join the ACA on an individual basis. Members also receive free of charge our quarterly magazine, *The Perchfisher*, and are able to attend meetings and fish-ins in various parts of the country.

Roger Miller – a member of 'The Perchfisher Group'.

A perch fisher's dream.

Projects covered by the group include a history of perch fishing, with details of all 3lb perch recorded in the UK press since 1870, and an investigation into the perch disease.

Index